I0038515

Designing Nanosensors for Chemical and Biological Applications

Nada F. Atta
Editor

Designing Nanosensors for Chemical and Biological Applications

International Frequency Sensor Association Publishing

Nada F. Atta, *Editor*
Designing Nanosensors for Chemical and Biological Applications

Copyright © 2017 by International Frequency Sensor Association (IFSA) Publishing

E-mail (for orders and customer service enquires): ifsa.books@sensorsportal.com

Visit our Home Page on http://www.sensorsportal.com

All rights reserved. This work may not be translated or copied in whole or in part without the written permission of the publisher (IFSA Publishing, S. L., Barcelona, Spain).

Neither the authors nor International Frequency Sensor Association Publishing accept any responsibility or liability for loss or damage occasioned to any person or property through using the material, instructions, methods or ideas contained herein, or acting or refraining from acting as a result of such use.

The use in this publication of trade names, trademarks, service marks, and similar terms, even if they are not identifies as such, is not to be taken as an expression of opinion as to whether or not they are subject to proprietary rights.

ISBN: 978-84-697-3290-8
BN-20170508-XX
BIC: TJF

Contents

Content.. 5
Contributors... 9
Preface 9

1. Liquid Crystal-nanostructures Hybrids and its Sensor Applications .. 17
  1.1. Introduction 17
  1.2. History of Liquid Crystals.. 19
  1.3. Classification of Liquid Crystals 21
    1.3.1. Thermotropic Liquid Crystal Mesophases 21
      1.3.1.1. Calamitic Mesogens.. 21
      1.3.1.2. Discotic Mesogens 26
      1.3.1.3. Bent-core Mesogens.. 27
    1.3.2. Lyotropic Liquid Crystal.. 27
  1.4. Characterization of Mesophases... 27
    1.4.1. Polarizing Optical Microscopy (POM)................................... 28
    1.4.2. Differential Scanning Calorimetry (DSC).............................. 30
    1.4.3. Powder X-ray Diffraction (PXRD)... 31
  1.5. Applications 32
    1.5.1. Uses of Liquid Crystals.. 32
      1.5.1.1. Liquid Crystal Display (LCD)... 32
      1.5.1.2. Liquid Crystal Thermometers... 33
      1.5.1.3. Liquid crystal Lasers.. 33
      1.5.1.4. Polymer Dispersed Liquid Crystal (PDLC) 33
      1.5.1.5. Soapy Water 34
      1.5.1.6. Electrochemical Modified Sensors 34
    1.5.2. Uses of Ionic Liquids.. 40
      1.5.2.1. Electrochemical Modified Sensor...................................... 40
      1.5.2.2. Drug Delivery (Biological Reactions Media)..................... 42
      1.5.2.3. Treatment of High-level Nuclear Waste 42
      1.5.2.4. Removing of Metal Ions.. 43
  1.6. Conclusion... 43
  References.. 44

2. Importance and Applications of Some Mediators in Chemistry........... 51
  2.1. Mediators and Electrochemistry 51
    2.1.1. Chemically Modified Electrodes 51
  2.2. Applications of Chemically Modified Electrodes.......................... 52
  2.3. Approaches to Chemically Modified Electrodes 53
  2.4. Examples of Some Important Modifiers.. 54
    2.4.1. Ferrocene.. 54
      2.4.1.1. Structure.. 54
      2.4.1.2. Bonding... 55
      2.4.1.3. Redox Chemistry 56

2.4.1.4. Applications of Ferrocene and its Derivatives *57*
2.4.2. Cyclodextrins. *62*
    2.4.2.1. Structure *62*
    2.4.2.2. Types of Cyclodextrins *63*
    2.4.2.3. Gamma-cyclodextrin *65*
    2.4.2.4. Beta- cyclodextrin. *66*
    2.4.2.5. Applications. *67*
    2.4.2.6. Analytical Techniques to Characterize Drug–CD Complexes. *72*
2.4.3. Phthalocyanine. *80*
    2.4.3.1. Properties. *80*
    2.4.3.2. Phthalocyanines Derivatives *81*
    2.4.3.3. Applications. *81*
    2.4.3.4. Examples of Metallophthalocyanine. *86*
2.5. Conclusion. 89
References 89

3. Self-Assembly Monolayers: New Strategy of Surface Modification
for Sensor Applications **95**
3.1. General Introduction. 95
    3.1.1. SAM Structure *96*
    3.1.2. Different Types of SAMs. *97*
        3.1.2.1. SAM of Organothiols. *97*
        3.1.2.2. SAMs of Phosphates *99*
        3.1.2.3. SAMs of Phthalocyanines *99*
        3.1.2.4. SAMs of Silanes *100*
        3.1.2.5. SAMs of Carbenes and Diazonium Salts *100*
    3.1.3. Different Types of Substrates for SAMs. *100*
        3.1.3.1. Gold, Platinum and Glassy Carbon. *101*
        3.1.3.2. Mercury. *102*
        3.1.3.3. Silver and Copper. *102*
        3.1.3.4. Nickel. *103*
        3.1.3.5. Palladium *103*
        3.1.3.6. Zinc. *104*
        3.1.3.7. Silicon. *105*
    3.1.4. Assembly Process of Alkanethiols Over Gold. *105*
3.2. Methods for SAMs Preparation 107
    3.2.1. Chemical Deposition of SAMs. *107*
        3.2.1.1. Solution Deposition. *107*
        3.2.1.2. Gas Phase Deposition *108*
    3.2.2. Electrochemical Deposition of SAMs *109*
3.3. Some Factors Related to the SAMs 109
    3.3.1. Effect of Chain Length. *109*
    3.3.2. Effect of Mixed SAMs *110*
    3.3.3. Effect of Number of S-atoms in the Thiol Compound *111*
    3.3.4. Effect of Time of Deposition in Chemical Growth Method. *112*
    3.3.5. Effect of Macro and Nano Au Substrate. *113*
3.4. Characterization of SAM Modified Electrodes. 115

3.4.1. Electrochemical Characterization *115*
   3.4.1.1. EIS *115*
   3.4.1.2. Electrochemical Desorption *115*
3.4.2. Surface Characterization *117*
   3.4.2.1. XPS *117*
   3.4.2.2. FT-IR *118*
   3.4.2.3. Raman Spectroscopy *118*
   3.4.2.4. SEM *119*
   3.4.2.5. TEM *119*
   3.4.2.6. AFM *120*
   3.4.2.7. NMR *121*
   3.4.2.8. UV-Vis *121*
   3.4.2.9. Contact Angle Goniometry *122*
3.5. Application of SAMs 122
   3.5.1. Metal Sensors *123*
   3.5.2. Neurotransmitter Sensors *124*
   3.5.3. Sensor for Pharmaceutically Important Drugs *129*
   3.5.4. Sensor for Ascorbic and Uric Acids *130*
   3.5.5. Immunosensor *131*
   3.5.6. DNA Sensor *132*
   3.5.7. Hydrogen Peroxide Sensors *135*
   3.5.8. Glucose Sensors *136*
   3.5.9. Cholesterol Sensors *139*
References 141

4. Perovskites: Smart Nanomaterials for Sensory Applications **149**
4.1. Introduction to Perovskite Oxides Nano-materials 149
4.2. Models for Perovskites 152
   4.2.1. The Ionic Models *152*
   4.2.2. Madelung and Electrostatic Potentials *152*
   4.2.3. Covalent Mixing *155*
4.3. General Characteristics of Perovskites 155
4.4. Doping of Perovskites 157
4.5. Some Examples of Perovskites 160
   4.5.1. Sr_2PdO_3 .... *160*
   4.5.2. $NdFeO_3$ *162*
4.6. Methods of Perovskite Synthesis 163
   4.6.1. Citrate–nitrate Combustion Methods *163*
   4.6.2. Co-precipitation Method *164*
   4.6.3. Microwave Synthesis *164*
4.7. Characterization of Perovskites 165
   4.7.1. XRD 166
   4.7.2. Scanning Electron Microscopy and Transmission Electron Microscopy
      (SEM, TEM) *168*
   4.7.3. Brunauer Emmett Teller Measurements (BET) *173*
   4.7.4. Thermal Analysis *173*
   4.7.5. FTIR174
   4.7.6. X-ray Photoelectron Spectroscopy (XPS) *175*

4.8. Applications of Perovskites 178

  4.8.1. Gas Sensor 179

  4.8.2. Glucose and H₂O₂ Sensor 180

  4.8.3. Neurotransmitters Sensor 184

  4.8.4. Sensor for Drugs 188

  4.8.5. Sensor for Amino Acids 192

References 193

5. Functionalized Carbon Based Materials for Sensing and Biosensing Applications: from Graphite to Graphene..207

5.1. Introduction 207

5.2. Graphite and Carbon Paste 208

  5.2.1. Polymeric Functionalization 209

  5.2.2. Metal and Metal Oxide Nanoparticles 212

  5.2.3. Mediator and Organic Modifier 214

5.3. Graphene 217

  5.3.1. Graphene Functionalization 218

  5.3.2. Hydrogenation and Halogenation of Graphene 219

  5.3.3. Hydroxylation of Graphene 221

  5.3.4. Carboxylation and Addition of Organic Groups 222

  5.3.5. Substitutional Doping with Foreign Atoms 230

5.4. Combined Carbon Materials 235

References 238

Index **255**

Contributors

Nada F. Atta

Nada F. Atta currently is a Professor of Chemistry at the Department of Chemistry, Faculty of Science, University of Cairo, Egypt. She earned a PhD degree from the University of Cincinnati, Ohio, USA. Current research interests are: new and advanced materials with emphasis on nano-structured materials, imprinted sol-gel materials, molecular recognition, nano-particles modified surfaces for Catalysis and sensors applications.

Ahmed Galal

Ahmed Galal is a Professor of Chemistry at the Department of Chemistry, Faculty of Science, University of Cairo, Egypt. He is currently Visiting Professor, at the University of Kuwait. Research interests are in the areas of electrochemical sensors, nano-materials, conducting Polymers, corrosion and passivity of metals and environmental chemistry. Other research interest is the use of chemically converted graphene in electrocatalysis and surface coating.

Ekram H. El-Ads

Ekram El-Ads is a Lecturer at the Department of Chemistry, Faculty of Science, University of Cairo, Egypt. She earned a PhD degree from the same university.

Hagar K. Hassan

Hagar Hassan is an Assistant Lecturer at the Department of Chemistry, Faculty of Science, University of Cairo, Egypt. She earned an MSc degree from the same university.

Shereen M. Azab

Shereen Azab is currently a Researcher at the National Organization for Drug Control and Research of Egypt. She earned a PhD degree from University of Cairo, Egypt.

Asmaa H. Ibrahim

Asmaa Ibrahim is currently a Researcher at the National Organization for Drug Control and Research of Egypt. She earned a PhD degree from University of Cairo, Egypt.

Preface

For the last two decades nanotechnology resulted in noticeable excitement among scientific and technological communities, and it is anticipated that nanotechnology will revolutionize the world in the future. A corner stone to this revolution is the development of novel nanomaterials. Nanotechnology is changing the world and the way we live, creating scientific advances and new products that are smaller, faster in delivery, effective, safer, and more reliable. There is a considerable potential for nanotechnology to open the door to the development of inexpensive, portable devices that can rapidly detect, identify, and quantify biological and chemical substances. In this respect, nanosensors are expected to lead to revolutionary applications, including early disease detection that can result in faster treatments and better outcomes, as well as the early and accurate detection of environmental pollutants, contaminants, and even biological or chemical weapons. Due to the diverse nature of these potential applications, nanosensors are expected to impact multiple sectors of the economy, including the healthcare, pharmaceutical, agricultural, food, environmental, consumer products, and defense sectors.

The present book aims at providing the readers with some of the most recent development of new and advanced materials and their applications as nanosensors. Examples of such materials are ferrocene and cyclodextrines as mediators, ionic liquid crystals, self-assembled monolayers on macro/ nano-structures, perovskite nanomaterials and functionalized carbon materials. The emphasis of the book will be devoted to the difference in properties and its relation to the mechanism of detection and specificity. Miniaturization on the other hand, is of unique importance for sensors applications. The chapters of this book present the usage of robust, small, sensitive and reliable sensors that take advantage of the growing interest in nano-structures. Different chemical species are taken as good example of the determination of different chemical substances industrially, medically and environmentally.

The book includes five chapters namely liquid crystal-nanostructures hybrids and its sensor applications, importance and applications of some mediators in chemistry, self-assembly monolayers: new strategy of surface modification for sensor applications, perovskites: smart nanomaterials for sensory applications and the last chapter about functionalized carbon based materials for sensing and biosensing applications: from graphite to graphene.

Chapter 1. Liquid crystals (LC) are sometimes referred to the fourth state of matter as they combine the anisotropic properties of the crystalline solid and the liquid-like nature of the isotropic liquid phase. Liquid crystal display (LCD) has proven to be one of the most important and profitable technological application of liquid crystal. Ionic liquid crystals (ILC) are soft, ordered materials that consist solely of ions. They are very promising candidates for use an anisotropic ion conductors in molecular electronics, batteries, fuel cells and capacitors. One of the most promising features of ionic liquid crystals is ionic conductivity and wide electrochemical potential windows which promoted their use as electrochemical modified sensors. An interesting feature is that these compounds mimic the natural bio-based ionic liquid crystals such as cell membranes structures because it exhibited excellent electrically conductivity property, hydrophilicity and large specific surface area. Since cell membranes are actually a form of naturally occurring ionic liquid crystals, therefore the main strategy developed for interaction of these membranes with the drug intake into cells is of prime importance. Therefore, the use of ILC as part of the electrode material under the influence of applied potential is inherent importance. Closely related to ionic liquid crystals are the so-called ionic liquids (IL). Room temperature ionic liquids (RTILs) have many specific physicochemical properties such as high chemical and thermal stability, relatively high ionic conductivity, negligible vapor pressure and wide electrochemical windows. RTILs have been widely used in the field of electrochemistry and electroanalysis.

Chapter 2. Mediation has extensively been used in sensor applications as it facilitates the exchange of charge between the sensing element and the chemical compound under investigation. Many biological compounds exhibit irreversible redox behavior as a result of slow heterogeneous electron transfer at electrode surfaces. In order to study the electrochemical behavior of these biocomponents, redox mediators are used to facilitate the electron transfer process. A redox mediator can be a hyper branched polymer having redox moieties incorporated into its structure and/or chemically attached to its periphery. It is attached to an electrode and assists in transferring electrons between the electrode and a redox enzyme thus allowing detection at lower potentials and reducing interferences. In addition, mediated biosensors offer other advantages such as increased linear responses and perhaps extended biosensor lifetime. A good redox mediator for a biosensor has to fulfill some characteristics such as: (1) an operating potential ideally ~0 V (versus SCE), where oxidation of most electrochemical interferents is avoided; (2) a fast reaction rate with the enzyme; (3) fast electron transfer kinetics; (4) no reaction with oxygen; (5) stable oxidized and reduced forms, etc. Typical mediators that are widely investigated included: ferrocene and its derivatives, cyclodextrines, and other metal hexacyanoferrates, quinones, metal phthalocyanines and methyl viologen. In this chapter the characteristics of some ideal mediators and structural information are discussed to provide a current survey of compounds having suitable redox mediation characteristics and applications.

Chapter 3. Self-Assembled Monolayer (SAM), an elegant way to orient as well as address electrically a molecular component of interest and is a precise modification of the surface structure on a nanometer-scale. This approach has opened up new era of exploration and has a profound impact on sensors and biosensors due to its unique properties. The stable, well-organized and densely packed SAMs formed by thiols on gold/ nanogold electrodes offer advantages such as selectivity, sensitivity, reproducibility, short response time and small overpotential in electrocatalytic reactions. In addition, gold nanoparticle modifications can greatly increase the immobilized amount of S-functionalized compounds and enhance the Au–S bond and stability of SAMs. On the other hand, the sensory applications at SAMs modified gold nanoparticle surface in presence of surfactants proved excellent and will be addressed in details. This chapter updates the reader with the scientific progress of the growth, structure and current applications of SAMs and highlights some of SAMs expected future applications.

Chapter 4. Inorganic perovskite-type oxides are fascinating nano-materials for wide applications in catalysis, fuel cells and electrochemical sensing. Perovskites are excellent materials that can enhance the catalytic performance and construct highly sensitive sensors therefore they are recently utilized in electrochemical sensing. They exhibit attractive physical and chemical characteristics such as electronic conductivity, electrically active structure, mobility of the oxide ions within the crystal, variations on the oxygen content and super-magnetic, photocatalytic, thermoelectric and dielectric properties. These nanomaterials present advantages over other types of materials in sensing applications such as high catalytic activity, selectivity, sensitivity, unique long term stability and excellent reproducibility. This chapter introduces a comprehensive coverage of the progress in perovskites research and its sensing application. Emphasis is given towards several intrinsic properties of the different class of perovskites namely electronic conductivity, electrically active structure and electrochemical performance in terms of synthesis routes and stability.

Chapter 5. Construction of new sensors based on the physical or chemical modifications of electrodes have attracted the attention of several researchers since the early work on electrochemical sensors. Carbon can be considered as the element of life due to its wide abundance and using; it is found everywhere, in air, in wood in meals, even in your furniture and clothes. So, it deserves to be the element of life worthily. Scientists knew this fact well, therefore for many years the carbon allotropes have been used in many applications including: energy conversion, energy storage, environmental aspects, field emission displays, electronics and sensors. Functionalization of the electrode materials is very crucial. One of the aims of the electrode functionalization is to enhance the electrocatalytic activity toward the analyte and in the same time to reduce the reactivity toward some other species which are called interfering materials. Additionally, the electrode functionalization may be used to increase the resistivity of the electrode surface toward surface fouling that result in enhancing the long-term stability of the electrode. So, we can conclude that; functionalization of electrodes increases the reactivity, stability and selectivity of the electrode material. One of the disadvantages of the carbon materials is its poor solubility that can be tuned by surface functionalization. Functionalization of carbon materials can be done by many routs some of them is wet and some is dry. In another words the carbon functionalization may be covalent or non- covalent. Covalent functionalization is based on the formation of a covalent bond between carbon and other components. Non-covalent functionalization is mainly based on adsorption via Van der Waals force, hydrogen bonds, electrostatic force and π-stacking interactions. Some of the functional entities that are usually used for carbon functionalization are: polymeric materials, biomolecules, nanoparticles or element doping, or by introducing an organic group to their surfaces which is usually oxygen-containing one. The scope of this chapter is to highlight up-to-date the using of the functionalized carbon materials including; graphite, activated carbon, carbon paste, carbon nanotube, fullerene and even graphene sheets, in sensing applications. We are going to discuss in details the methods of carbon functionalization and their characterization that have been mentioned in literature. Also, the purpose behind the combination between two different carbon allotropes and their using as a combined electrode surface for sensing applications are illustrated in this chapter.

1.

Liquid Crystal-nanostructures Hybrids and its Sensor Applications

Nada F. Atta and Asmaa H. Ibrahim

1.1. Introduction

Ionic liquids (ILs) have attracted increasing interest lately in several areas such as chemistry, physics, engineering, material science, molecular biochemistry, energy and fuels, among others. Furthermore, the range of ILs used has been broadened and there has been a significant increase in the scope of both physical and chemical ILs properties [1, 2]. Ionic liquids are the salts having very low melting temperature. Ionic liquids have received great interests recently because of their unusual properties as liquids. These unique properties of ionic liquids have already been mentioned in some books, so we do not repeat them here more than simply summarize them in Table 1.1.

Table 1.1. Basic Characteristics of Ionic Liquids.

Low melting point	-Treated as liquid at ambient temperature -Wide usable temperature range
Non-volatility	-Thermal stability -Non-flammability
Composed by ions	-High ion density -High ion conductivity
Organic ions	-Various kinds of salts -Designable -Unlimited combination

The most important properties of electrolyte solutions are non-volatility and high ion conductivity. Because ionic liquids are composed of only ions, they show very high ionic conductivity, non-volatility, and non-flammability. The non-flammable liquids with high ionic conductivity are practical materials for use in electrochemistry. At the same time, the non-flammability and non-volatility inherent in ion conductive liquids open new possibilities in other fields as well. Because most energy devices can accidentally explode or ignite, for motor vehicles there is plenty of incentive to seek safe materials. Ionic liquids are being developed for energy devices. It is therefore important to have an understanding of the basic properties of these interesting materials. The ionic liquids are multi-purpose materials, so there should be considerable (and unexpected) applications. Closely related to ionic liquids are the so-called ionic liquid crystals.

Ionic liquid crystals (ILCs) are matter in a state that has properties between those of conventional liquid and those of solid crystal [3]. For instance, a liquid crystal may flow like a liquid, but its molecules may be oriented in a crystal-like way. There are many different types of liquid-crystal phases which can be distinguished by their different optical properties (such as birefringence). When viewed under a microscope using a polarized light source, different liquid crystal phases will appear. LC materials may not always be in a liquid-crystal phase (just as water may turn into ice or steam). Ionic liquid crystals are soft and ordered materials that consist solely of ions.

This work focuses on the liquid crystalline state (also mesophase or mesomorphic state) that can be found in between the solid and liquid state and combines the properties of both [4-11]. Liquid crystals can be considered as anisotropic liquids as they appear soft and fluid, but at the same time possess anisotropic physical properties. That is, their refractive index, dielectric permittivity, magnetic susceptibility, mechanical properties, etc. depend on the direction in which they are measured. Mesophase formation is attributed to the relaxation of the crystal lattice forces between the molecules in the solid state. During this process a certain degree of orientational (and in some cases positional) ordering is maintained, unlike the transition to a fully disordered isotropic liquid. On a molecular level, the order that is preserved in the liquid-crystalline phase is a consequence of the orientational alignment of interacting molecules (Fig. 1.1).

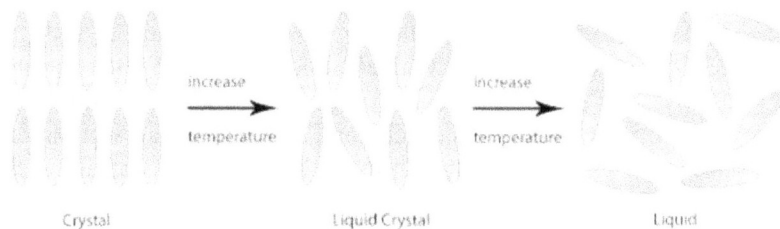

Crystal   Liquid Crystal   Liquid

Fig. 1.1. Schematic representation of the gradual loss of molecular order in a thermotropic liquid crystal upon heating, reused with permission.

1.2. History of Liquid Crystals

Liquid-crystallinity was first observed by Friedrich Reinitzer in 1888 [12]. Reinitzer, who was in fact a botanist, noticed a reversible colour change during multiple heating and cooling cycles of cholesterol derivatives. In addition, he found that these compounds exhibited a peculiar melting behaviour. Cholesteryl benzoate (Fig. 1.2) didn't show one but two melting points.

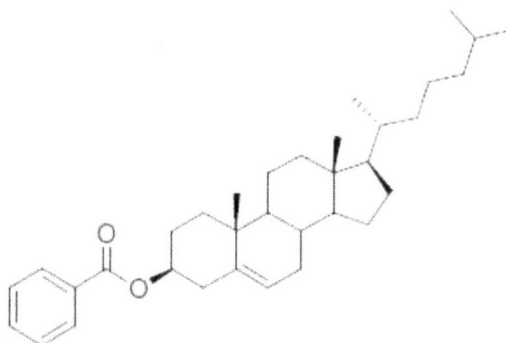

Fig. 1.2. Cholesteryl benzoate, reused with permission.

Entangled by his observations, he contacted Otto Lehmann, a crystallographer from Aachen who was in possession of a polarizing microscope. Only one year later, in 1889, Lehmann launched the concept of 'flowing crystals' [13]. This idea was quickly picked up by contemporary scientists Gattermann and Ritschke, who reported on azoxyphenol ethers that turned into cloudy fluids upon melting (originally assuming that it concerned a mixture) [14]. They were the

19

first to denote the phenomenon Flüssige Kristalle or liquid crystals. At the same time, however, Georg Quincke and Theodor Wulff contested the existence of pure substances possessing both liquid-like and crystalline properties, stating that Lehmann's liquid crystals were in fact colloidal suspensions and that the observed birefringence was due to the solid component [15]. This point of view was shared by Tammann and Nernst who referred to colloidal emulsions (drop of liquid suspended in another liquid) rather than suspensions [16-18]. In 1904 the colloidal hypothesis was contradicted by an electrophoresis experiment of Coehn, and separately also Bredig and Schukowsky, in which colloidal particles could be removed from the emulsion [19, 20]. Physicochemical measurements performed by Schenk showed no discontinuities in viscosity and density, again a proof of purity [21]. In the meantime a wide range of organic substances that showed liquid-crystalline behaviour was synthesized, of which a great deal in the group of Halle. In 1906 Vorländer published the first review on the subject in the highly rated Berichte der Deutschen Chemischen Gesellschaft, followed by an undoubtedly even more interesting article in which he stated: 'All results obtained so far suggest that the crystalline-liquid state results from a molecular structure which is as linear as possible. The birefringence of the crystalline-liquid state is caused by the anisotropy of the molecule [22, 23]. This was clearly a turning point as the origin of this anisotropic phase was now directly related to the molecular shape! After World War I George Friedel introduced a new nomenclature because in his opinion the term 'liquid crystal' was greatly misleading. In his manuscript 'Les états mésomorphes de la matière' he distinguished smectics (smegma = soap), nematics (nema = thread), and cholesterics (often formed by cholesterol derivatives) [24]. He recognized that smectic materials were layered structures and proposed X-ray measurements in order to verify such a lamellar organization. In 1923, his son Edmond Friedel and Maurice de Broglie confirmed the existence of equidistant molecular layers in smectic materials [25]. Although a large amount of liquid-crystalline molecules had been synthesized by then and a hint of understanding the molecular ordering within different phases was set, there was no thorough physical background for this phenomenon. An elaborate discussion on the topic was held at the first symposium on Liquid Crystals and Anisotropic Melts in 1933[26]. The swarm theory proposed by Emil Bose as early as 1908 was the subject of dispute having its supporters (Ornstein, Kast) and opponents (Zocher, Osein) [27]. An important absentee at the conference was Professor V.K. Freedericksz who set the basis of modern display technology with experiments on the magnetic alignment of liquid crystals [28].

1.3. Classification of Liquid Crystals

Liquid crystals are classified into thermotropic and lyotropic liquid crystals.

1.3.1. Thermotropic Liquid Crystal Mesophases

Thermotropic mesophases are induced by heating a pure solid (mesomorphic) compound. The added heat causes an increase in thermal motion of the molecules, so that the three-dimensional order present in the crystal lattice is partially broken. The transition from the solid state to a mesomorphic state takes place at the melting temperature Tm. Upon further heating, the kinetic energy of the molecules becomes too large to preserve strong anisometric interactions, characteristic for the liquid-crystalline state, resulting in a fully disordered, isotropic phase. This transition is associated with a clearing temperature Tc. Mesomorphic materials can show a sequence of mesophases before reaching the isotropisation point; this is referred to as poly(meso)morphism. Transitions between different mesophases and between a mesophase and the isotropic liquid state are in general reversible, and they occur with little hysteresis in temperature as opposed to the crystallization process, which is often subject to supercooling. For some compounds liquid-crystallinity is only observed on cooling due to a greater stability of the crystalline phase compared to the respective mesophase. A so-called monotropic phase is always observed below the melting point. Enantiotropic mesophases are formed on heating as well as on cooling.

For thermotropic systems that occur in a certain temperature range, thermotropic mesophases can be classified accordingly to the molecular shape into rod-like molecules (calamitic mesogens), disk-like molecules (discotic mesogens) and V-shaped molecules (bananashaped or bent-core mesogens).

1.3.1.1. Calamitic Mesogens

Rod-like molecules usually form nematic or smectic phases. These mesogens generally consist of a rigid core to which flexible chains are attached (Fig. 1.3).

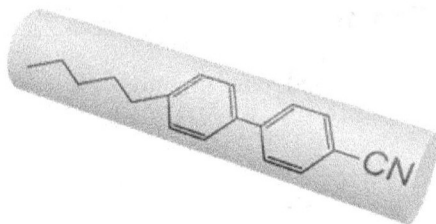

Fig. 1.3. Molecular structure of a rod-like liquid crystal,
reused with permission.

On heating the aliphatic tails melt providing the fluidity present in a
liquid crystal. Stronger interactions such as ionic interactions between
the rigid cores ensure orientational alignment and thus the anisotropy of
the mesophase. Phases formed by calamitic mesogens have a defined
thermodynamic order. If, hypothetically, one single compound would
exhibit all phases described below, it would (on cooling) go through a
phase sequence of [29]:

$$I {\rightarrow} N {\rightarrow} SmA {\rightarrow} SmC {\rightarrow} SmB {\rightarrow} SmI {\rightarrow} SmF {\rightarrow} Cr$$

This phase sequence is correlated to the amount of order present in one
mesophase relative to another. Less ordered phases are situated at the
high-temperature end, whereas additional order is introduced on
lowering the temperature.

1.3.1.1.1. Nematic phase (N)

The nematic phase is the least ordered mesophase of all calamitic
mesophases. The long molecular axes of the rod-like molecules are on
average aligned in one preferential direction (along the director ñ)
(Fig. 1.4). The molecules are able to rotate around their long axes and
possess three-dimensional translational freedom. The molecular
orientational distribution around the director is expressed using the
orientational order parameter S:

$$S = \frac{1}{2} \langle 3cos^2\theta - 1 \rangle,$$

where θ is the angle between the long molecular axis of an individual
molecule and the director.

Fig. 1.4. Schematic representation of the alignment in the nematic phase, reused with permission.

In a nematic phase, the calamitic or rod-shaped organic molecules have no positional order, but they self-align to have long-range directional order with their long axes roughly parallel [30]. Thus, the molecules are free to flow and their center of mass positions are randomly distributed as in a liquid, but still maintain their long-range directional order. Although nematic phases are optically uniaxial, Freiser theoretically predicted a biaxial nematic phase N in 1970. In this phase the two directions perpendicular to the director ñ are no longer equivalent and the degree of rotational freedom of the molecules about the long molecular axes is restricted. The biaxial nematic phase received a great deal of attention over the last couple of years, not only because of the difficulties in obtaining this phase, but also because low molecular weight molecules that form the Nb phase are expected to show a much faster switching rate than uniaxial nematics [31, 32]. Finally, it is worth noting that nematic phases are rarely observed for ionic liquid crystals, although some examples exist [33-35].

1.3.1.1.2. Smectic phase (Sm)

The smectic phases, which are found at lower temperatures than the nematic, form well defined layers that can slide over one another in a manner similar to that of soap. The word "smectic" originates from the Latin word "smecticus", meaning cleaning, or having soap like properties [36]. The smectics are thus positionally ordered along one direction. In the Smectic A phase (Fig. 1.5), the molecules are oriented along the layer normal, while in the Smectic C phase they are tilted away from the layer normal. These phases are liquid-like within the layers. There are many different smectic phases, all characterized by different types and degrees of positional and orientational order [37, 38].

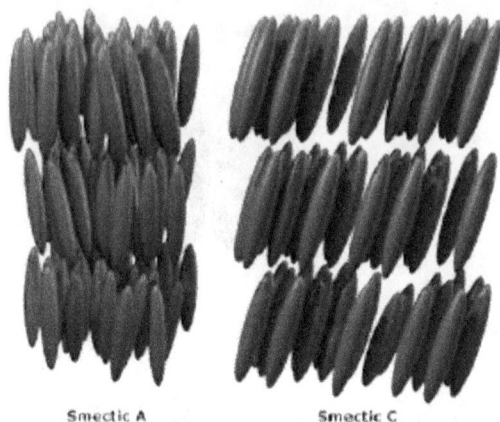

Smectic A   Smectic C

Fig. 1.5. Schematic representation of the molecular ordering in the smectic A, and C phase, reused with permission.

The smectic B, smectic I and smectic F phases are classified as hexatic phases because the molecules adopt a hexagonal ordering within the smectic layers. Although the hexatic phases are characterized by a short-range positional ordering, the hexagonal net stretches over a long distance. So, while the smectic layers can still slide over each other, the hexagons will be oriented in the same way. Such ordering is referred to as two-dimensional bond-orientational order. The smectic B phase is an orthogonal hexatic phase, whereas the smectic I and smectic F phases are tilted modifications. The latter two are distinguished by the tilt direction of the molecules: towards an apex of the hexagon for the smectic I phase and towards an edge in the case of a smectic F phase.

1.3.1.1.3. Chiral Phases

The chiral nematic phase exhibits chirality. This phase is often called the cholesteric phase (Fig. 1.6) because it was first observed for cholesterol derivatives. Only chiral molecules (i.e., those that have no internal planes of symmetry) can give rise to such a phase. This phase exhibits a twisting of the molecules perpendicular to the director, with the molecular axis parallel to the director. The finite twist angle between adjacent molecules is due to their asymmetric packing, which results in longer-range chiral order. In the smectic C phase (an asterisk denotes a chiral phase), the molecules have positional ordering in a layered structure (as in the other smectic phases), with the molecules tilted by a finite angle with respect to the layer normal. The chirality induces a finite azimuthal twist from one layer to the next, producing a spiral twisting of the molecular axis along the layer normal [37, 39, 40].

Fig. 1.6. Schematic of ordering in chiral liquid crystal phases. The chiral nematic phase (left), also called the cholesteric phase, and the smectic C phase (right), reused with permission.

In the cholesteric phase 'planes' of nematic assemblies associate in a fashion so that the directors of consecutive planes are twisted in a helix, resulting in left- or right-handed supramolecular structures (Fig. 1.7). Comparable to other helical structures the chiral nematic phase is characterized by a pitch P, defined as the length over which the director turns 360°. The optical activity of chiral nematogens is highly dependent on the length of the pitch P. The plane of linearly polarized light will be rotated by a cholesteric sample if the pitch is much longer than the wavelength of the incident light. However, when the pitch length is comparable to the wavelength of the incident light, circular dichroism is observed. Circularly polarized light of the same handedness as the helix will be reflected selectively, according to the law of Bragg, which results in an iridescent appearance. It should be kept in mind that chiral these optical properties can only be observed macroscopically if the optical axes are aligned perpendicular to the working substrate. In addition, the helical pitch is temperature-dependent so that small changes in the pitch length (or temperature) will cause colour changes. As such thermochromic cholesterics have been used as liquid-crystal temperature sensors in thermographic studies [41].

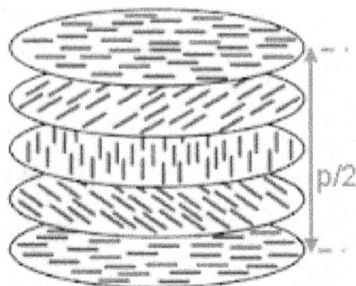

Fig. 1.7. Schematic representation of the molecular ordering in the chiral nematic phase, p refers to the chiral pitch, reused with permission.

1.3.1.2. Discotic Mesogens

Disk-shaped LC molecules can orient themselves in a layer-like fashion known as the discotic nematic phase (Fig. 1.8). If the disks pack into stacks, the phase is called a discotic columnar. The columns themselves may be organized into rectangular or hexagonal arrays. Chiral discotic phases are also known chiral nematic phase.

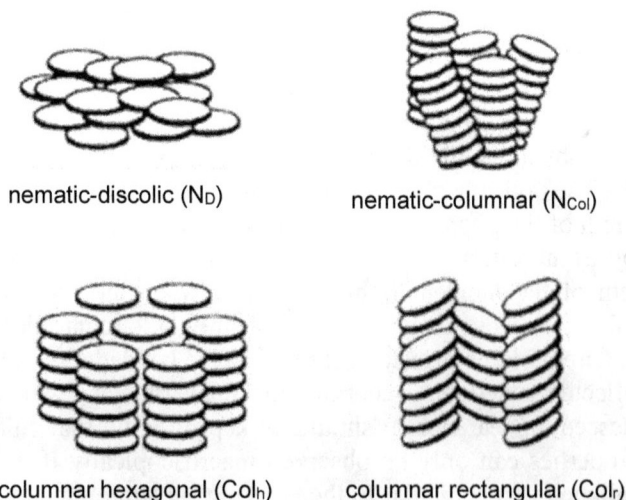

nematic-discolic (N_D) nematic-columnar (N_{Col})

columnar hexagonal (Col_h) columnar rectangular (Col_r)

Fig. 1.8. Schematic representation of the molecular ordering in the discotic nematic (N phase (ND), nematic columnar phase (N_{Col}) and hexagonal columnar phase (Col_h), reused with permission.

1.3.1.3. Bent-core Mesogens

Bent-core or banana-shaped liquid crystals have attracted considerable attention due to their unique mesomorphic properties [42-44]. Unlike calamitic and discotic molecules, bent shaped mesogens have restricted rotational freedom along their main molecular axes because of steric effects. Bearing this in mind, bent-core molecules were thought to be unsuitable for liquid crystal design. Nevertheless, since a report on the ferroelectric switching of 1,3-phenylene bis[4-(4-n-octyl-oxy-phenyl-imino-methyl)-benzoate by Niori et al. [45], numerous papers have been published on the topic.

1.3.2. Lyotropic Liquid Crystal

A lyotropic liquid crystal consists of two or more components that exhibit liquid crystalline properties in certain concentration ranges (Fig. 1.9). In the lyotropic phases, solvent molecules fill the space around the compounds to provide fluidity to the system [46]. In contrast to thermotropic liquid crystals, these lyotropics have another degree of freedom of concentration that enables them to induce a variety of different phases. A compound that has two immiscible hydrophilic and hydrophobic parts within the same molecule is called an amphiphilic molecule. Many amphiphilic molecules show lyotropic liquid-crystalline phase sequences depending on the volume balances between the hydrophilic part and hydrophobic part. These structures are formed through the micro-phase segregation of two incompatible components on a nanometer scale. Soap is an everyday example of a lyotropic liquid crystal.

1.4. Characterization of Mesophases

Mesophase characterization generally involves the use of three complementary techniques. The identification of the type of mesophase that is formed at a certain temperature is done by means of polarizing optical microscopy (POM), whereas differential scanning calorimetry (DSC) provides accurate information on the transition temperatures and order of the transition. Nevertheless, the most powerful technique to get a thorough understanding of the packing of the molecules in the liquid-crystalline state is powder X-ray diffraction (PXRD).

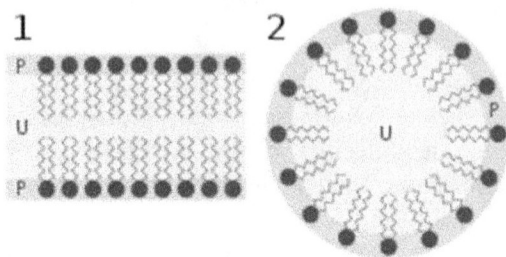

Fig. 1.9. Structure of lyotropic liquid crystal. The red heads of surfactant molecules are in contact with water, whereas the tails are immersed in oil (blue): bilayer (left) and micelle (right), reused with permission.

1.4.1. Polarizing Optical Microscopy (POM)

When light passes from one medium (medium 1) through another (medium 2), its velocity and propagation direction will change. This phenomenon, commonly known as refraction, is uniform in all orientations for isotropic fluids and follows Snell's law:

$$\frac{sin\theta_1}{sin\theta_2} = \frac{v_1}{v_2} = \frac{n_2}{n_1},$$

where θ_1 is the angle of incidence, θ_2 is the angle of refraction, v_1 and v_2 are the velocities and n_1, n_2 are the refractive indices of the respective media. However, anisotropic media such as liquid crystals and crystalline solids are birefringent which means that the value of the refractive index is directionally dependent. For uniaxial mesophases (nematic, smectic A, etc.) incident light will be split up into two components: an ordinary ray, that obeys Snell's law of refraction and an extraordinary ray that does not obeys Snell's law. Both beams will propagate in different directions and at different velocities which results in a phase difference δ, given by:

$$\delta = \frac{2\pi}{\lambda}(n_e - n_0)d$$

With λ the vacuum wavelength, ne and no different indices of refraction and d the distance travelled in the medium. Now, in polarizing optical microscopy the incident light is linearly polarized by a polariser (Fig. 1.10).

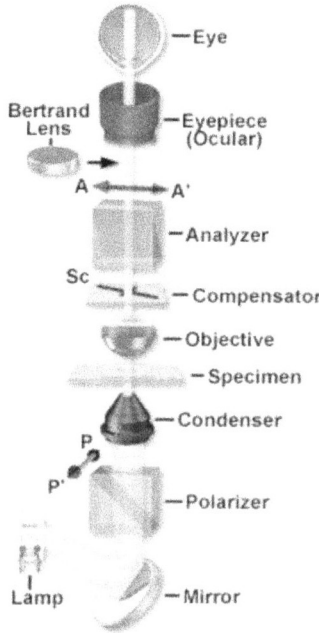

Fig. 1.10. Typical experimental setup of a polarising microscope: the sample is placed on a heating stage that controls the temperature (here designated as specimen). The polarisers are situated at both sides of the sample and are generally crossed with respect to each other, reused with permission.

On passing through a birefringent sample the light will be converted to elliptically polarized light of which one component is transmitted by the analyzer. The intensity I of the outcoming light can be calculated from the following equation:

$$I = I_0 sin^2 2\varphi sin^2 \frac{\delta}{2},$$

where I_0 is the light intensity after the polarizer, φ is the azimuthal angle (the angle between the analyzer and the projection of the optical axis onto the sample plane) and δ is the phase difference. According to the last equation that the transmittance is zero either if the azimuthal angle φ equals to 0° or 90° or if the phase difference δ is equal to zero. The former ($\varphi = 0°$ or 90°) is conform with a planar alignment for which the optical axis is parallel to one of the polarizer directions whereas the latter ($\delta = 0°$) corresponds to a uniform homeotropic orientation. In both cases a black or pseudo-isotropic image is observed.

The property of birefringence, characteristic for liquid identification of the type of mesophase as different optical images (textures) are observed for different molecular organizations. In fact, a texture is the result of alterations from a uniform director orientation and is caused by various kinds of defects present in the sample. Such defects can also be found in the solid, crystalline state, but in that case they usually have submicroscopic dimensions. In Schlieren texture, which is typical for the nematic phase, the best images are obtained by cooling from the isotropic liquid. On further cooling, below the melting point, a crystalline solid is formed that can be distinguished from crystalline state by its lack of fluidity.

The study of miscibilities is another useful way for the classification of different liquid-crystalline phases. In general, binary systems of liquid crystals can be assigned to the same phase type if they are miscible. However, it is possible that, due to large structural differences between the compounds, two liquid-crystalline phases of the same type do not mix [47].

1.4.2. Differential Scanning Calorimetry (DSC)

Differential scanning calorimetry (DSC) is used to accurately determine the transition temperatures at which a phase change occurs. At the transition from a crystalline solid to a liquid-crystalline or isotropic phase (or between any two phases) a loss of order will take place, corresponding to an entropy change. In DSC the enthalpy change, which is directly related to the change in entropy ($\Delta S = \Delta H/T$), accompanied by such a phase transition is measured. A typical DSC measuring system is built up by a twin-type design. That is, either temperature variations or an exchange of heat between a sample pan and an (empty) reference pan is measured. The measured signal is then proportional to the heat flow rate Φ, resulting in Φ (t) curve.

Depending on the method used to determine the heat flow rate two techniques are distinguished: heat-flux DSC and power-compensating DSC. In heat-flux DSC the reference pan and sample pan are placed in one single furnace and heated at the same rate, usually 10 °C/min, by the heating plate. At a transition, the sample will be at a lower temperature than the reference. From the temperature lag and the known sample mass the enthalpy change can then be calculated [48].

In power-compensating DSC the reference pan and sample pan are placed in two separate furnaces. Both furnaces are kept at the same temperature and the power to compensate for the temperature difference at a phase transition is measured.

Differential scanning calorimetry cannot be used to identify a mesophase, but one can get an idea of the type of phase transition. Phase transitions can be first order or second order. A first order transition is characterized by a discontinuous jump in the first derivative of the free Gibbs energy ΔG ($\partial G/\partial T$). First order transitions, like Cr-N transition, can be recognized by sharp peaks and large enthalpy changes. The enthalpy changes vary depending on the loss of order between two phases (for example: the enthalpy changes of Cr-SmA transitions are around 30-50 kJ/mol, while SmA-N transitions are accompanied with enthalpy changes around 1 kJ/mol. A second order transition is characterized by a continuous first derivative of the free Gibbs energy, and a discontinuous second derivative of the free Gibbs energy. In this case the heat capacity Cp will exhibit a discontinuous jump. Second order transitions like SmC-SmA can be recognized as a bump rather than a peak in a DSC thermogram.

1.4.3. Powder X-ray Diffraction (PXRD)

High-temperature X-ray diffraction allows the unambiguous identification of the type of mesophase and is used to investigate the molecular organization within the mesophase. Diffraction is a phenomenon that comes about when radiation is elastically scattered by the electron clouds of a material scatter [49]. Constructive and destructive interference of the X-rays lead to a diffraction pattern that is characteristic for each type of mesophase.

For a set of planes Bragg's law describes the angles for which constructive interference occurs:

$$n\lambda = 2d \sin\theta,$$

where n is an integer, λ is the wavelength (usually 1.54 Å), d is the distance between to atom planes and θ is the angle of incidence.

The technique used for the identification of a liquid-crystalline phase is powder X-ray diffraction. Here, the sample is heated into the mesophase giving rise to differently oriented monodomains. The intensity I of the scattered X-rays is measured as a function of the scattering angle (2θ, typically between $0°$ and $40°$), leading to a diffraction pattern containing several diffraction signals. From the diffractogram detailed information on the molecular ordering in the mesophase can be obtained:

(I) The layer spacing d which is related to the position of the peaks.

(II) The long-distance ordering within the mesophase can be derived from the sharpness of the peaks.

(III) The number of the diffracting atoms which is proportional to the intensity of the peaks.

(IV) The phase type (nematic, smectic, columnar, cubic) determines the ratio of the peak positions at small angles.

(V) The translational ordering over short distances (for example the lateral order in the smectic phases) which is related to the wide-angle peaks.

1.5. Applications

1.5.1. Uses of Liquid Crystals

Liquids crystals have attracted increasing interest lately in several areas such as chemistry, physics, engineering, material science, molecular biochemistry, energy and fuels, among others. Furthermore, the range of LCs used has been broadened, and there has been a significant increase in the scope of both physical and chemical LCs' properties. LCs are defined as liquid organic salts composed entirely of ions, and a melting point criterion has been proposed to distinguish between molten salts and ionic liquids.

1.5.1.1. Liquid Crystal Display (LCD)

Liquid crystals find wide use in liquid crystal displays, which rely on the optical properties of certain liquid crystalline substances in the presence or absence of an electric field. In a typical device, a liquid crystal layer (typically 10 μm thick) sits between two polarizers that are crossed (oriented at $90°$ to one another). The liquid crystal alignment is chosen so that its relaxed phase is a twisted one. This twisted phase reorients light that has passed through the first polarizer, allowing its transmission through the second polarizer (and reflected back to the observer if a reflector is provided). The device thus appears transparent. When an electric field is applied to the LC layer, the long molecular axes tend to align parallel to the electric field thus gradually untwisting in the center of the liquid crystal layer. In this state, the LC molecules do not reorient light, so the light polarized at the first polarizer is absorbed at the second polarizer, and the device loses transparency with increasing voltage. In this way, the electric field can be used to make a pixel switch between transparent or opaque on command. Color LCD systems use the same technique, with color filters used to generate red, green, and blue pixels [50]. Similar principles can be used to make other liquid crystal based optical devices [51]. Phones, monitors, TVs and GPS are examples of liquid crystal displays.

1.5.1.2. Liquid Crystal Thermometers

Thermotropic chiral LCs whose pitch varies strongly with temperature can be used as crude liquid crystal thermometers, since the color of the material will change as the pitch is changed. Liquid crystal color transitions are used on many aquarium and pool thermometers as well as on thermometers for infants or baths [52]. Other liquid crystal materials change color when stretched or stressed. Thus, liquid crystal sheets are often used in industry to look for hot spots, map heat flow, measure stress distribution patterns, and so on. Liquid crystal in fluid form is used to detect electrically generated hot spots for failure analysis in the semiconductor industry [53].

1.5.1.3. Liquid crystal Lasers

Liquid crystal lasers use a liquid crystal in the lasing medium as a distributed feedback mechanism instead of external mirrors. Emission at a photonic band gap created by the periodic dielectric structure of the liquid crystal gives a low-threshold high-output device with stable monochromatic emission [54, 55].

1.5.1.4. Polymer Dispersed Liquid Crystal (PDLC)

Polymer Dispersed Liquid Crystal (PDLC) sheets and rolls are available as adhesive backed Smart film which can be applied to windows and electrically switched between transparent and opaque to provide privacy.

1.5.1.5. Soapy Water

Many common fluids, such as soapy water, are in fact liquid crystals. Soap forms a variety of LC phases depending on its concentration in water [56].

1.5.1.6. Electrochemical Modified Sensors

1.5.1.6.1. Ionic Liquid Crystal Modified Carbon Paste Electrode

For the first time, we use ionic liquid crystal namely (1-Butyl-1-methylpiperidinium hexafluorophosphate) to successfully fabricate modified carbon paste in presence of anionic surfactant (sodium dodecyl sulfate) [ILCMCPE.....SDS] for the electrochemical determination of drugs such as Benazepril HCl (BN) [57] and Enoxacin (EN) [58]. The results using ILCMCPE.....SDS was compared to other ionic liquid modified carbon paste electrodes namely (1-n-Hexyl-3-methyl imidazolium tetrafluoroborate) (ILMCPE1.....SDS) and (1-Butyl-4-methyl pyridinium tetrafluoroborate) (ILMCPE2.....SDS).

Fig. 1.11 shows Cyclic voltammograms (CV) of 1.0×10^{-3} molL^{-1} of BN in B-R buffer pH 7.4 at scan rate 100 mVs^{-1} recorded at different working electrodes, CPE.....SDS, ILMCPE1.....SDS, ILMCPE2.....SDS and ILCMCPE.....SDS. For ILCMCPE.....SDS, the anodic peak current 92.16 µA at 0.73 V was obtained which is six times higher than that obtained at bare CPE and it is more higher with negative shift in the oxidation potential value than in case of ILMCPE.....SDS. Some ionic materials are known also to form amphitropic liquid crystals; one of the particularly interesting properties of these materials being their ease of deprotonation whereas the positive charge is localized on the nitrogen atom of piperidinium salts which lead to wider electrochemical potential window than for imidazolium and pyridinium salts at which the positive charge is delocalized over the aromatic ring, also the solid state structure of the ionic liquid crystal helps in the formation of ordered films [59, 60]. Furthermore this remarkable enhancement might be caused by the synergism effect of preconcentration/accumulation of SDS with high polarizability and ionic conductivity of the ionic liquid crystal, which provides a remarkable increase in the rate of electron transfer process [61, 62]. Also for enoxacin (EN), Fig. 1.12 shows that ILCMCPE...SDS, the anodic peak current was 70.5 µA at 0.97 V which is more than five times higher than that obtained at bare CPE and gave more negative shift in the oxidation potential value than in case of ILMCPE. Fig. 1.13 shows the ionic liquid crystals mimic the natural bio-based ionic liquid crystals such as cell membranes structures in their interactions with drugs.

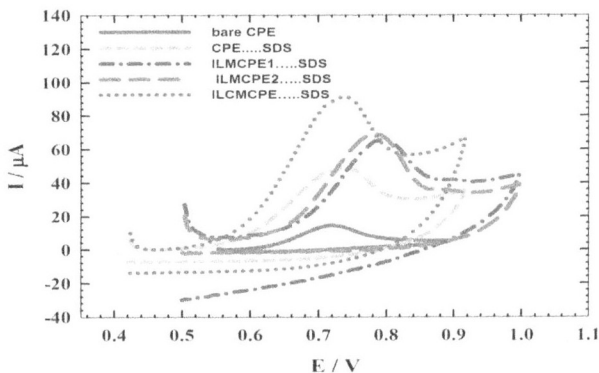

Fig. 1.11. Cyclic voltammograms (CV) of 1.0×10^{-3} molL^{-1} of BN in B-R buffer pH 7.4 at scan rate 100 mVs^{-1} recorded at different working electrodes (bare CP (solid line), CPE.....SDS (small dashed line), ILMCPE1.....SDS (dashed dotted line), ILMCPE2..... SDS (large dashed line) and ILCMCPE.....SDS (dotted line), reused with permission.

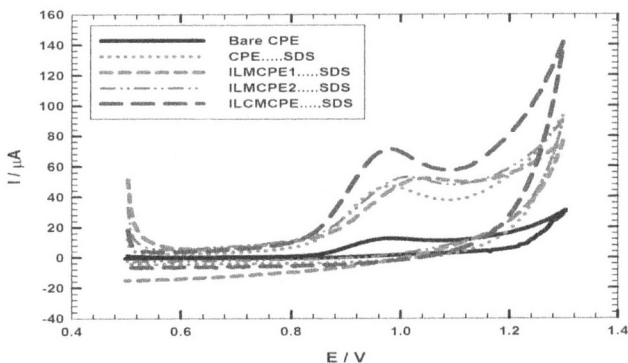

Fig. 1.12. Cyclic voltammograms (CV) of 1.0×10^{-3} mol L^{-1} of EN in B-R buffer pH 7.4 at scan rate 100 mVs^{-1} recorded at different working electrodes (bare CP (solid line), CPE.....SDS (small dashed line), ILMCPE1.....SDS (dashed dotted line), ILMCPE2..... SDS (large dashed line) and ILCMCPE.....SDS (dotted line), reused with permission.

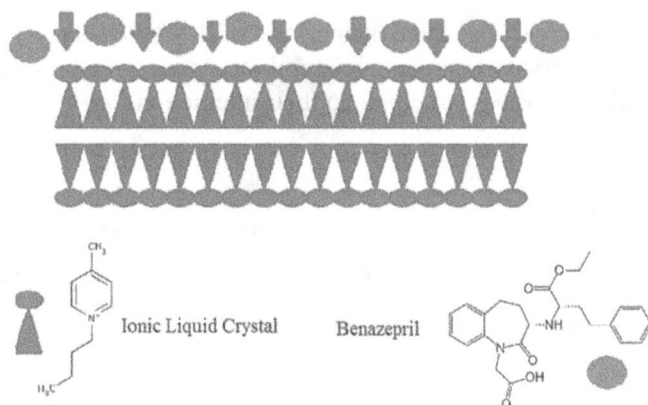

Fig. 1.13. Schematic diagram of ILC modified CP, reused with permission.

1.5.1.6.2. NiONps / Ionic Liquid Crystal Modified Carbon Paste Electrode

The ionic liquid crystal 1-butyl-1-methylpiperidinium hexafluorophosphate and nickel oxide nanoparticles were used to construct a carbon composite electrode. This novel composite was used successfully as a sensor platform for the determination of paracetamol (ACOP) and some neurotransmitters such as dopamine (DA), levodopa (L-Dopa), norepinephrine (NEP) and serotonin (ST) [63]. Several advantages are realized in this approach due to the unique properties of nanomaterials and ionic liquid crystals, and the ease of fabrication of the carbon composite electrode. The modified sensor was evaluated and compared with nickel oxide nanoparticles/ionic liquid modified electrodes in the presence of surfactants. Modification with ionic liquid crystals showed superior current signals compared to ionic liquids. The interaction of surfactants with neurotransmitters resulted in preconcentration of the drug at the ionic liquid crystal interface that allowed both ionic channeling and charge transfer mediation.

Fig. 1.14 shows Cyclic voltammograms (CV) of 1.0×10^{-3} mol L^{-1} of ACOP in B-R buffer pH 7.4 at scan rate 100 mVs^{-1} recorded at different working electrodes bare CP, bare CPE..SDS, NiONps /ILMCPE1…SDS, NiO-Nps/ILMCPE…SDS and NiO-Nps/ILCMCPE…SDS. For NiO-Nps/ILCMCPE.....SDS modified surface; the anodic peak current 115 µA at 0.425 V was obtained which is nearly three times higher than that obtained at bare CPE and it is much higher with negative shift in the oxidation potential value than in case of

NiONps/ILMCPE reflecting the synergistic effect achieved by the components of the proposed sensor.

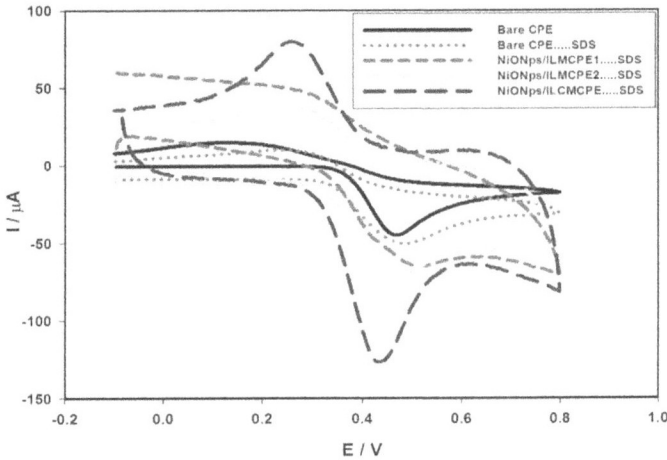

Fig. 1.14. Cyclic voltammograms (CV) of 1.0×10^{-3} mol L^{-1} of ACOP in B-R buffer pH 7.4 at scan rate 100 mVs^{-1} recorded at different working electrodes (bare CP (solid line), bare CPE.....SDS(dotted line), NiONps/ILMCPE1.....SDS (small dashed dotted line), NiONps/ILMCPE2.....SDS(dashed dotted line) and NiONps/ILCMCPE.....SDS(large dashed line), reused with permission.

Fig. 1.15 shows Cyclic voltammograms (CV) of 1.0×10^{-3} mol L^{-1} of L-Dopa in B-R buffer pH 7.4 at scan rate 100 mVs^{-1} recorded at different working electrodes namely bare CP, bare CPE.....SDS, NiONps/ILMCPE1...SDS, NiONps/ILMCPE2...SDS and NiO-Nps/ ILCMCPE...SDS. The NiONps/ILCMCPE...SDS has the highest anodic peak current with more negative oxidation potential than other electrodes. This is attributed to the intrinsic characteristics of the components of the proposed sensor. Also, this result proved the fundamental criteria of the ionic liquid crystals in comparison with other ionic liquids.

The electrode was successfully employed for simultaneous determination of paracetamol and neurotransmitters. As shown in Fig. 1.16A, ACOP exhibits well-defined differential pulse voltammograms DPV with good separations from DA in B-R buffer (pH 7.4) by changing the concentration of both ACOP (4.4→ 133.0 μmol L^{-1}) and DA (4.4→36.0 μmol L^{-1}). The current responses due to the oxidation of DA (at 170 mV) and ACOP (at 374 mV) with a peak separation of 204 mV were observed. With increasing their concentrations, the current responses of both ACOP and DA were increased linearly with a correlation coefficient of 0.9996 and 0.9981, respectively, also the regression equation for ACOP was found to be: $Ip(\mu A) = 0.247 \, c(\mu mol \, L^{-1}) + 4.951$, while the regression equation for DA was: $Ip(\mu A) = 0.832 \, c(\mu mol \, L^{-1}) + 5.748$.

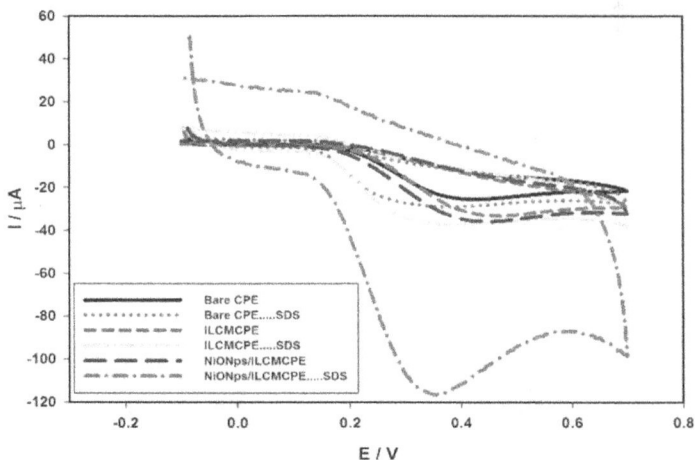

Fig. 1.15. Cyclic voltammograms (CV) of 1.0×10^{-3} mol L^{-1} of L-Dopa in B-R buffer pH 7.4 at scan rate 100 mVs^{-1} recorded at different working electrodes (bare CP (solid line), bare CPE.....SDS(dotted line), NiONps/ILMCPE1.....SDS (small dashed dotted line), NiONps/ILMCPE2.....SDS(dashe dotted line) and NiONps/ILCMCPE.....SDS(large dashed line), reused with permission.

Simultaneous determination of ACOP and DA in the mixture was also investigated when the concentration of one species changed, whereas the other was kept constant. Fig. 1.16B shows that the peak current of ACOP increased with an increase in the ACOP concentration ($8.89 \rightarrow 89.0$ μmol L^{-1}) while the concentration of DA was kept constant (4.4 μmol L^{-1}). Also keeping the concentration of ACOP constant (4.0 μmol L^{-1}), the oxidation peak current of DA was positively proportional to its concentration ($4.44 \rightarrow 44.11$ μmol L^{-1}) (Fig. 1.16 C). It should be noted that, the change of concentration of one compound did not have significant influence on the peak current and peak potential of the other compound.

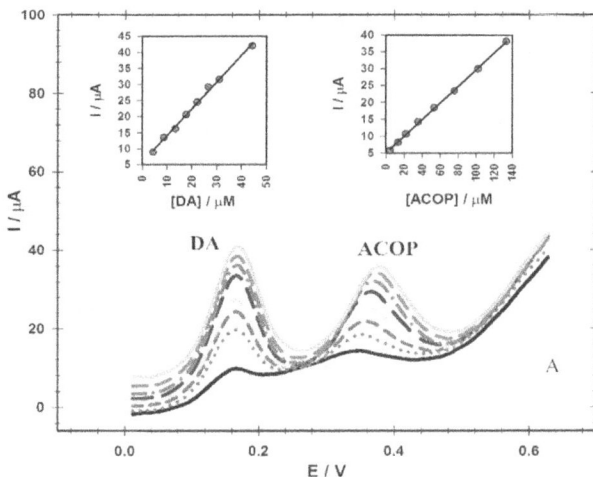

Fig. 1.16A. The differential pulse voltammograms obtained with NiONps/ ILCMCPE…..SDS with good separations between ACOP and DA by changing the concentration of both ACOP (4.4→133.0 µmol L⁻¹) and DA (4.4→36.0 µmol L⁻¹) in 0.04 M B-R buffer pH 7.4, scan rate 10 mV/s, reused with permission.

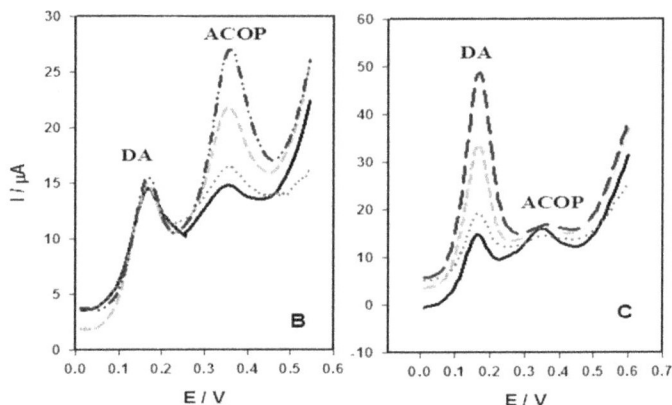

Fig. 1.16 B, C. The differential pulse voltammograms obtained with NiONps/ILCMCPE…..SDS with an increase in the ACOP concentration (8.89→89.0 µmol L⁻¹) while the concentration of DA was kept constant (4.4 µmol L⁻¹)) in 0.04 M B-R buffer pH 7.4, scan rate 10 mV/s. 23C) The differential pulse voltammograms obtained with NiONps/ ILCMCPE…..SDS with increase in the DA concentration (4.44→44.11 µmol L⁻¹) while the concentration of ACOP was kept constant (4.0 µmol L⁻¹)) in 0.04 M B-R buffer pH 7.4, scan rate 10 mV/s, reused with permission.

The calibration curve was investigated using NiONps/ ILCMCPE.....SDS which has the linear dynamic range of 44.4×10^{-7} - 3.33×10^{-5} mol L^{-1} for paracetamol sensing with a correlation coefficient of 0.999 and a limit of detection of 8.61×10^{-9} mol L^{-1}. The electrode was successfully employed for the direct determination of paracetamol in human urine samples and for paracetamol assay in pharmaceutical formulations. High reproducibility and selectivity in the presence of potential interfering species were ascertained for this electrode. The good properties of this modified electrode will expand its application in electrochemical field for the determination of other drugs in biological fluids without any interference.

1.5.2. Uses of Ionic Liquids

Ionic liquids are very popular materials and they enjoy a plethora of applications in various domains of physical sciences.

For example, they are used as "solvents" for organic, organometallic syntheses and catalysis, electrolytes in electrochemistry, lubricants in fuel and solar cells, as a stationary phase for chromatography, as matrices for mass spectrometry, supports for the immobilization of enzymes, templates for the synthesis of mesoporous, nano-materials and ordered films, materials for embalming and tissue preservation, etc. Fig. 1.17 is provided a map of the applications of ionic liquids.

1.5.2.1. Electrochemical Modified Sensor

The most important properties of ionic liquids are: thermal stability, low vapour pressure, electric conductivity, liquid crystal structures, high electro-elasticity, high heat capacity, large electrochemical window and good electrochemical stability. These properties make ILs suitable for their use in different electrochemical applications such as biosensors. The major application of these biosensors so far is in neurotransmitters (NT) sensing because of their important role in the central nervous system. Dopamine (DA) is a neurotransmitter with a very important role in the central nervous system. The DA determination in patients with Parkinson disease is important, because low levels of DA increased the symptomatic effects [64–66]. Ionic liquid (n-octyl pyridinium hexafluorophosphate [OPy]PF$_6$) modified carbon paste electrode (ILCPE) has been used for investigation of 1 mM dopamine solution in buffer PBS, pH = 7.4 using cyclic voltammetric (CV) technique at scan rate 100 mv/s.

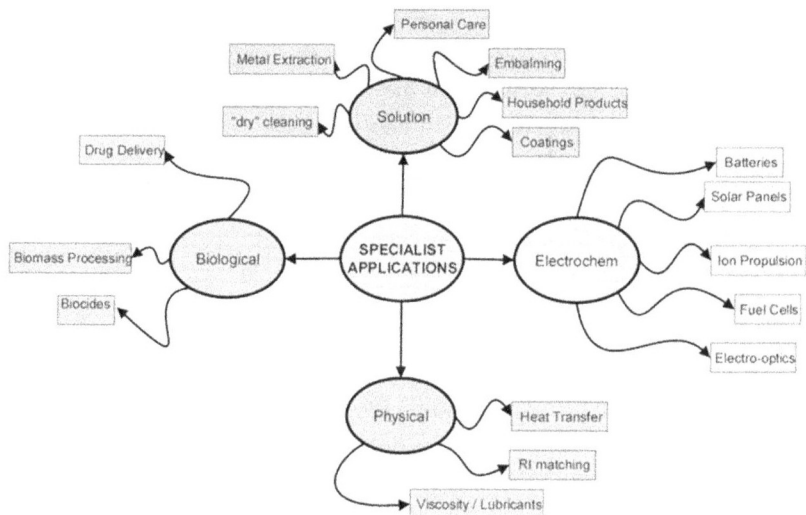

Fig. 1.17. A map of applications of ionic liquids, reused with permission.

The anodic peak current (Ipa) observed in the detection of dopamine was obtained with [OPy]CPE is higher than that of CPE, which suggests a best performance with this electrode in the presence of DA. Opallo et al. [67] described these phenomena as a result of a sum of currents from the reaction of DA with the carbon directly and through the ionic liquid/carbon interface, due to the ionic conductivity of the ionic liquid which is not appreciated in the CPE case.

The catecholamines are a group of compounds bearing a dihydroxyphenyl moiety [68] known as main neurotransmitters, and have been employed as markers of neuroblastoma, stress condition and other autonomic nervous system disorders [69]. Epinephrine (EP), also known as adrenaline, one of the important catecholamines, plays a central role during physical or mental stress and also stimulates a series of actions of the sympathetic nervous system (SNS) known as the "flight or fight response" [70]. Low levels of EP have been found in patients with Parkinson's disease [71–72]. From the point of view of medicine, it is a drug for emergency treatment in severe allergic reaction, cardiac arrest and sepsis [73]. Ionic liquid modified carbon nanotubes paste electrode (IL/CNTPE) has been used to determine 300 μM EP using CV at phosphate buffer (PBS) pH 7.0. The electrochemistry of EP was studied at four different electrodes; bare carbon paste electrode (CPE), carbon nanotubes paste electrode (CNTPE), ionic liquid carbon paste electrode (IL/CPE) and ionic liquid carbon nanotubes paste electrode (IL/CNTPE) [74]. The results indicated that the presence of CNTs on IL surface had great improvement on the electrochemical response, which was partly due to excellent characteristics of CNTs such as good electrical conductivity, high chemical stability and high surface area.

1.5.2.2. Drug Delivery (Biological Reactions Media)

As enzymes and proteins are stable in ionic liquids, this may open the possibility for ionic liquids to be used in biological reactions, such as the synthesis of pharmaceuticals [75]. Work by Fujita et al [76, 77] and others [78, 79] have shown that some proteins are, in fact, soluble, stable and remain active in some ILs. As a case study Cytochrome c (cyt. c) was found to have enhanced solubility and stability in a biocompatible IL solution based on the dihydrogen phosphate anion [77]. This is an important observation since proteins are sometimes unstable when handled in vitro, and stabilizing agents are a necessary component to ensure their long-term stability. This is especially true of proteins that have pharmaceutical potential since lack of stability is a limitation to widespread use of some protein therapeutics. It has been well documented that enzyme performance in an IL is affected by several parameters including water activity, pH and impurities [80]. Other important factors that play a role in enzyme stability / activity include IL polarity, hydrogen bond basicity and nucleophillicity of anions, ion kosmotropocity and viscosity. Although outside the scope of this discussion, these areas have been discussed in an excellent review on the topic by Zhao [81]. Abe et al [82] recently synthesized a number of phosphonium salts that have an alkyl ether group present. The phosphonium salts moiety is commonly found in living creatures, and it was hypothesized that this family of ILs have good affinity with enzyme proteins and may provide a good environment for enzymes.

1.5.2.3. Treatment of High-level Nuclear Waste

Ionizing radiation does not affect ionic liquids, so they could even be used to treat high-level nuclear waste [83]. For example, Ionic liquids (ILs) have been widely considered as potential green solvents for CO_2

capture [84]. This is related to the hydrodynamics of CO_2 absorption with ionic liquids.

1.5.2.4. Removing of Metal Ions

In another application, Davis and Rogers have designed and synthesized several new ionic liquids to remove cadmium and mercury from contaminated water. When these water-insoluble ionic liquids come in contact with contaminated water, they snatch the metal ions out of water and sequester them in the ionic liquid [85].

1.6. Conclusion

This chapter introduces some basic properties of ionic liquids and liquid crystals. Ionic liquids (ILs) have attracted increasing interest lately in several areas such as chemistry, physics, engineering, material science, molecular biochemistry, energy and fuels. Due to significant increase in the scope of both physical and chemical properties of IL, they can be used as:

- Solvents for organic and organometallic syntheses.

- Catalysis and electrolytes in electrochemistry.

- Lubricants in fuel and solar cells.

- Stationary phase for chromatography.

- Matrices for mass spectrometry.

- Supports for the immobilization of enzymes.

- Templates for the synthesis of mesoporous materials for embalming and tissue preservation.

On the other hand, Ionic liquid crystals (ILCs) are matter in a state that has properties between those of conventional liquid and those of solid crystal. ILCs are soft, ordered materials that consist solely of ions. Liquid crystalline state can be known as mesophase or mesomorphic state. Polarizing optical microscopy (POM), Differential scanning calorimetry (DSC) and Powder x-ray diffraction (PXRD) are techniques used for characterization of liquid crystal mesophase. Because of their properties high ionic conductivity, wide potential window, non-volatility, and non-flammability, liquid crystals have attracted increasing interest lately in several areas such as chemistry, physics, engineering, material science, and molecular biochemistry. They have many applications such as: liquid crystal display (LCD), liquid crystal thermometers, liquid crystal lasers, polymer dispersed liquid crystal (PDLC) and soapy water.

References

[1]. D. R. MacFarlane, K. R. Seddon, Ionic Liquids- Progress on the Fundamental Issues, *Australian Journal of Chemistry,* Vol. 60, Issue 1, 2007, pp. 3-5.

[2]. T. Torimoto, T. Tsuda, K. Okazaki, S. Kuwabata, New Frontiers in Materials Science Opened by Ionic Liquids, *Advanced Materials*, Vol. 22, Issue 11, 2010, pp. 1196-1221.

[3]. O. Lehmann, Über fliessende Krystalle, *Zeitschrift für Physikalische Chemie*, Vol. 4, 1889, pp. 462–72.

[4]. P. J. Collings, J. S. Patel, Eds., Handbook of Liquid Crystal Research, *Oxford University Press,* Oxford, New York, 1997.

[5]. P. J. Collings, M. Hird, Introduction to Liquid Crystals: Chemistry and Physics, *Taylor & Francis Ltd.,* London, 1997.

[6]. P. J. Collings, Liquid Crystals: Nature's Delicate Phase of Matter, 2nd ed., *Princeton University Press,* Princeton, 2001.

[7]. D. Demus, J. W. Goodby, G. W. Gray, H.-W. Spiess, V. Vill, Eds., Handbook of Liquid Crystals. Volume 1: Fundamentals, *Wiley-VCH,* Weinheim, 1998.

[8]. D. Demus, J. W. Goodby, G. W. Gray, H.-W. Spiess, V. Vill, Eds., Handbook of Liquid Crystals. Volume 2A: Low Molecular Weight Liquid Crystals I, *Wiley-VCH,* Weinheim, 1998.

[9]. D. Demus, J. W. Goodby, G. W. Gray, H.-W. Spiess, V. Vill, Eds., Handbook of Liquid Crystals. Volume 2B: Low Molecular Weight Liquid Crystals II, *Wiley-VCH,* Weinheim, 1998.

[10]. D. Demus, J. W. Goodby, G. W. Gray, H.-W. Spiess, V. Vill, Eds., Handbook of Liquid Crystals. Volume 3: High Molecular Weight Liquid Crystals, *Wiley-VCH,* Weinheim, 1998.

[11]. I. W. Hamley, Introduction to Soft Matter, *John Wiley & Sons, Ltd.,* Chichester, 2000.

[12]. F. Reinitzer, Beiträge zur Kenntniss des Cholesterins, *Monatshefte fur Chemie*, Vol. 9, 1888, pp. 421-441.

[13]. O. Z. Lehmann, Über fliessende Krystalle, *Physical Chemistry,* Vol. 4, 1889, pp. 462-472.

[14]. L. Gattermann, A. Ritschke, Ueber Azoxyphenetoläther (On Azoxyphenol Ethers), *Berichte der deutschen chemischen Gesellschaft*, Vol. 23, 1890, pp. 1738-1750.

[15]. G. Quincke, Ueber freiwillige Bildung von hohlen Blasen, Schaum und Myelinformen durch ölsaure Alkalien und verwandte Etrscheinungen besonders des Protoplasmas, *Wiedemann's Annalen*, Vol. 53, 1894, pp. 593-631.

[16]. G. Tammann, Ueber die sogennanten flüssigen Krystalle, *Annalen der Physik*, Vol. 4, 1901, pp. 524-530.

[17]. G. Tammann, Ueber die sogennanten flüssigen Krystalle II, *Annalen der Physik*, Vol. 8, 1902, pp. 103-108.

[18]. G. Tammann, über die Natur der, flüssigen Kristalle, III, *Annalen der Physik*, Vol. 19, 1906, pp. 421-425.

[19]. G. Bredig, N. Schukowsky, Prüfung der Natur der flüssigen Krystalle mittels elektrischer Kataphorese, *Berichte der deutschen chemischen Gesellschaft*, Vol. 37, 1904, pp. 3419-3425.

[20]. G. Coehn, Über Flüssige Kristalle, *Zeit Elektrochemie*, Vol. 10, 1904, pp. 856-857.

[21]. R. Schenck, On crystalline liquids and liquid crystals, *Zeit Elektrochemie*, Vol. 11, 1905, pp. 951-955.

[22]. D. Vorländer, Ueber krystallinisch-flüssige Substanzen, *Berichte der deutschen chemischen Gesellschaft*, Vol. 39, 1906, pp. 803-810.

[23]. D. Vorländer, Einfluß der molekularen Gestalt auf den krystallinisch-flüssigen Zustand, *Berichte der deutschen chemischen Gesellschaft*, Vol. 40, 1907, pp. 1970-1972.

[24]. G. Friedel, Les Étates Mésomorphes de la Matière, *Annales de Physique (Paris)*, Vol. 18, 1922, pp. 273-474.

[25]. M. de Broglie, E. Friedel, X-Ray Diffraction by Smectic Materials, *Comptes rendus de l'Acamédie des Sciences*, Vol. 176, 1923, pp. 738-740.

[26]. Bernal, et al., Liquid crystals and anisotropic melts: A general discussion, *Transactions of the Faraday Society*, Vol. 29, 1933, pp. 1060-1085.

[27]. E. Bose, Zur theorie der anisotropen flusseigkeiten, *Physikalische Zeitschrift*, Vol. 9, 1908, pp. 708-713.

[28]. Freedericksz V., Zolina V., On the use of a magnetic field in the measurement of the forces tending to orient an anisotropic liquid in a thin homogeneous layer, *Transactions of the American Electrochemical Society*, Vol. 55, 1929, pp. 85-96.

[29]. F. Camerel, G. Ulrich, J. Barbera, R. Ziessel, Ionic Self-Assembly of Ammonium-Based Amphiphiles and Negatively Charged Bodipy and Porphyrin Luminophores, *Chemistry- A European Journal*, Vol. 13, 2007, pp. 2189- 2200.

[30]. James A. Rego, Jamie A. A. Harvey, Andrew L. MacKinnon, Elysse Gatdula, *Liquid Crystals*, Vol. 37, Issue 1, 2010, pp. 37 -43.

[31]. C. Tschierske, D. J. Photinos, Biaxial nematic phases, *Journal of Materials Chemistry*, Vol. 20, 2010, pp. 4263-4294.

[32]. D. W. Bruce, Towards the Biaxial nematic phase through molecular design, *The Chemical Record*, Vol. 4, 2004, pp. 10-22.

[33]. L. Lu, N. Sharma, G. A. N. Gowda, C. L. Khetrapal, R. G. Weiss, Enantiotropic Nematic Phases of Quaternary Ammonium Halide Salts Based on Trioctadecylamine, *Liquid Crystals*, Vol. 22, 1997, pp. 23-28.

[34]. F. Artzner, M. Veber, M. Clerc, A. M. Levelut, Evidence of nematic, hexagonal and rectangular columnar phases in thermotropic ionic liquid crystals, *Liquid Crystals*, Vol. 23, 1997, pp. 27-33.

[35]. D. W. Bruce, D. Dunmur, P. Maitlis, P. Styring, M. Esteruelas, L. Oro, M. Ros, J. L. Serrano, E. Sola, Nematic phases in ionic melts: mesogenic ionic complexes of silver(I), *Chemistry of Materials*, Vol. 1, 1989, pp. 479-481.

[36]. S. Chandrasekhar, Liquid Crystals (2nd ed.), Cambridge, *Cambridge University Press*, 1992.

[37]. P. G. de Gennes, J. Prost, The Physics of Liquid Crystals, *Clarendon Press*, Oxford, 1993.

[38]. 'Smectic', Merriam-Webster Dictionary.

[39]. I. Dierking, Textures of Liquid Crystals, *Wiley-VCH,* Weinheim, 2003.

[40]. P. J. Collings, M. Hird, Introduction to Liquid Crystals, *Taylor & Francis*, Bristol, PA, 1997.

[41]. I. Sage, Thermochromic liquid crystals, *Liquid Crystals*, Vol. 38, 2011, pp. 1551-1561.

[42]. G. Pelzl, S. Diele, W. Weissflog, Banana-shaped compounds: A new field of liquid crystals, *Advanced Materials*, Vol. 11, 1999, pp. 707-724.

[43]. M. B. Ros, J. L. Serrano, M. R. de la Fuente, C. L. Folcia, Banana-shaped liquid crystals: a new field to explore, *Journal of Materials Chemistry*, Vol. 15, 2005, pp. 5093-5098.

[44]. H. Takezoe, Y. Takanishi, Bent-core liquid crystals: Their mysterious and attractive world, *Japanese Journal of Applied Physics Part 1*, Vol. 45, 2006, pp. 597-625.

[45]. T. Niori, T. Sekine, J. Watanabe, T. Furukawa, H. Takezoe, Distinct ferroelectric smectic liquid crystals consisting of banana shaped achiral molecules, *Journal of Materials Chemistry*, Vol. 6, Issue 7, 1996, pp. 1231-1233.

[46]. Q. Liang, P. Liu, C. Liu, X. Jian, D. Hong, Y. Li., Synthesis and Properties of Lyotropic Liquid Crystalline Copolyamides Containing Phthalazinone Moieties and Ether Linkages, *Polymer*, Vol. 46, Issue 16, 2005, pp. 6258–6265.

[47]. D. Demus, L. Richter, Textures of Liquid Crystals, *Verlag Chemie,* Weinheim, New York, 1978.

[48]. G. Höhne, W. Hemminger, H. J. Flammersheim, Differential Scanning, Calorimetry: An Introduction for Practitioners, Second ed., *Springer*, New York, 2003.

[49]. Handbook of Liquid Crystals. Volume 1: Fundamentals, *Wiley-VCH,* Weinheim, 1998.

[50]. J. A. Castellano, Liquid Gold: The Story of Liquid Crystal Displays and the Creation of an Industry, *World Scientific Publishing,* 2005.

[51]. T. T. Alkeskjold, L. Scolari, D. Noordegraaf, J. Lægsgaard, J. Weirich, L. Wei, G. Tartarini, P. Bassi, S. Gauza, S. T. Wu, A. Bjarklev, Integrating liquid crystal based optical devices in photonic crystal, *Optical and Quantum Electronics,* Vol. 39, Issue 12, 2007, pp. 1009-1019.

[52]. R. G. Plimpton, Pool thermometer, *U. S. Patent 4,738,549,* Issued on April 19, 1988.

[53], Hot-spot detection techniques for ICs, acceleratedanalysis.com. Retrieved May 5, 2009.

[54]. V. I. Kopp, B. Fan, H. K. M. Vithana, A. Z. Genack, Low-threshold lasing at the edge of a photonic stop band in cholesteric liquid crystals, *Optics Express,* Vol. 23, Issue 21, 1998, pp. 1707–1709.

[55]. K. Dolgaleva, K. H. W. Simon, G. L. Svetlana, H. C. Shaw, S. Katie, W. B. Robert, Enhanced laser performance of cholesteric liquid crystals doped with oligofluorene dye, *Journal of the Optical Society of America,* Vol. 25, Issue 9, 2008, pp. 1496–1504.

[56]. V. Luzzati, H. Mustacchi, A. Skoulios, Structure of the Liquid-Crystal Phases of the Soap–water System: Middle Soap and Neat Soap, *Nature,* Vol. 180, Issue 4586, 1957, pp. 600-601.

[57]. N. F. Atta, A. Galal, S. M. Azab, A. H. Ibrahim, Electroanalysis of Benazepril Hydrochloride Antihypertensive Drug Using an Ionic Liquid Crystal Modified Carbon Paste Electrode, *Electroanalysis,* Vol. 27, 2015, pp. 1282-1292.

[58]. N. F. Atta, A. Galal, S. M. Azab, A. H. Ibrahim, Electrochemical Sensor based on Ionic Liquid Crystal Modified Carbon Paste Electrode in Presence of Surface Active Agents for Enoxacin Antibacterial Drug, *Journal of the Electrochemical Society,* Vol. 162, 2015, pp. B9-B15.

[59]. M. Shukla and S. Saha, A Comparative Study of Piperidinium and Imidazolium Based Ionic Liquids: Thermal, Spectroscopic and Theoretical Studies, Ch. 3, Department of Chemistry, Faculty of Science, *Banaras Hindu University,* Varanasi, India, 2013.

[60]. K. Lava, Ionic liquid crystals based on novel heterocyclic cores, Dissertation Katholieke Universiteit Leuven, Groep Wetenschap & Technologie, *Arenberg Doctoraatsschool*, Belgium, 2012.

[61]. N. Maleki, A. Safavi, F. Tajabadi, High-performance carbon composite electrode based on an ionic liquid as a binder, *Analytical Chemistry,* 2006, Vol. 78, Issue 11, 2006, pp. 3820–3826.

[62]. D. Wei, A. Ivaska, Applications of ionic liquids in electrochemical sensors, *Analytica Chimica Acta,* Vol. 607, Issue 2, 2008, pp. 126 –135.

[63]. N. F. Atta, A. H. Ibrahim, A. Galal, Nickel oxide nanoparticles / ionic liquid crystal modified carbon composite electrode for determination of neurotransmitters and paracetamol, *New Journal of Chemistry,* Vol. 40, 2016, pp. 662-673.

[64]. L. Tierney, S. McPhee, M. Papadakis, Diagnóstico Clínico y Tratamiento, Cap. 24, 25 y 26, Quinta Edic. Edit. El Manual Moderno, 2000, pp. 965-967, 1015, 1016, 1065.

[65]. A. Abdel-Baki, C. Ouetllet-Plamondon, A. Malla, Pharmacotherapy challenges in patients with first-episode psychosis, *Journal of Affective Disorders*, Vol. 138, 2012, S3-14.

[66]. D. Santos-García, M. Prieto-Formoso, R. de la Fuente-Fernández, Levodopa dosage determines adherence to long-acting dopamine agonists in Parkinson's disease, *Journal of Neurological Science*, Vol. 318, 2012, pp. 91-93.

[67]. M. Opallo, L. Lesniewski, A review on electrodes modified with ionic liquids, *Journal of Electroanalytical Chemistry*, Vol. 656, 2011, pp. 2-16.

[68]. H. K. Yildrim, A. Üren, U. Yücel, Food Technology, *Biotechnology*, Vol. 45, 2007, pp. 62–68.

[69]. S. G. Ball, I. G. Gunn, I. H. Doglus, Renal handling of dopa, norepinephrine, and epinephrine in the dog, *American Journal of Physiology, Renal Physiology*, Vol. 242, 1982, pp. F56–F62.

[70]. D. L. Wong, T. C. Tai, D. C. Wong-Faull, R. Claycomb, R. Kvetnansky, Adrenergic Responses to Stress, *Annals of the New York Academy of Sciences*, Vol. 1148, 2008, pp. 249-256.

[71]. C. W. Hsu, M. C. Yang, Electrochemical epinephrine sensor using artificial receptor synthesized by sol-gel process, *Sensors and Actuators B: Chemical*, Vol. 134, 2008, pp. 680-686.

[72]. L. I. B. Silva, F. D. P. Ferreira, A. C. Freitas, T. A. P. Rocha-Santos, A. C. Duarteb, Optical fiber biosensor coupled to chromatographic separation for screening of dopamine, norepinephrine and epinephrine in human urine and plasma, *Talanta*, Vol. 80, 2009, pp. 853-857.

[73]. C. G. Amorim, A. N. Araujo, M. C. B. S. M. Montenegro, Exploiting sequential injection analysis with lab-on-valve and miniaturized potentiometric detection, Epinephrine determination in pharmaceutical products, *Talanta*, Vol. 72, Issue 4, 2007, pp. 1255-1260.

[74]. T. Tavana, M. A. Khalilzadeh, H. Karimi-Maleh, A. A. Ensafi, H. Beitollahi, D. Zareyee, Sensitive voltammetric determination of epinephrine in the presence of acetaminophen at a novel ionic liquid modified carbon nanotubes paste electrode, *Journal of Molecular Liquids*, Vol. 168, 2012, pp. 69–74.

[75]. J. H. Davis, P. A. Fox, From Curiosities to Commodities: Ionic Liquids Begin the Transition, *Chemical Communications*, Issue 11, 2003, pp. 1209-1212.

[76]. H. Ohno, C. Suzuki, K. Fukumoto, M. Yoshizawa and K. Fujita, Electron Transfer Process of Poly(ethylene oxide)-Modified Cytochrome c in Imidazolium Type Ionic Liquid, *Chemistry Letters*, Vol. 32, 2003, pp. 450-451.

[77]. K. Fujita, D. R. MacFarlane and M. Forsyth, Protein solubilising and stabilising ionic liquids, *Chemical Communications*, Issue 38, 2005, pp. 4804-4806.

[78]. S. N. Baker, T. M. McCleskey, S. Pandey and G. A. Baker, Fluorescence studies of protein thermostability in ionic liquids, *Chemical Communications*, Issue 8, 2004, pp. 940-941.

[79]. J. A. Laszlo and D. L. Compton, Comparison of peroxidase activities of hemin, cytochrome c and microperoxidase-11 in molecular solvents and imidazolium-based ionic liquids, *Journal of Molecular Catalysis B: Enzymatic*, Vol. 18, 2002, pp. 109-120.

[80]. Z. Yang and W. Pan, Ionic liquids: Green solvents for nonaqueous biocatalysis, *Enzyme and Microbial Technology*, Vol. 37, 2005, pp. 19-28.

[81]. H. Zhao, Methods for stabilizing and activating enzymes in ionic liquids: A review, *Journal of Chemical Technology & Biotechnology*, Vol. 85, 2010, pp. 891-907.

[82]. Y. Abe, K. Yoshiyama, Y. Yagi, S. Hayase, M. Kawatsura and T. Itoh, A rational design of phosphonium salt type ionic liquids for ionic liquid coated-lipase catalyzed reaction, *Green Chemistry*, Vol. 12, 2010, pp. 1976-1980.

[83]. K. R. Seddon, A. Stark, M. J. Torres, Influence of chloride, water and organic solvents on the physical properties of ionic liquids, Queen's University Belfast, Béal Feirste, Northern Ireland, *United Kingdom Pure and Applied Chemistry*, Vol. 72, Issue 12, 2000, pp. 2275-2287.

[84]. X. Zhang, H. Dong, Y. Huang, C. Li, X. Zhang, Experimental study on gas holdup and bubble behavior in carbon capture systems with ionic liquid, *Chemical Engineering Journal*, Vol. 209, 2012, pp. 607–615.

[85]. A. E. Visser, R. P. Swatloski, W. M. Reichert, R. Mayton, S. Sheff, A. Wierzbicki, J. H. Davis, R. D. Rogers, Task-Specific Ionic Liquids Incorporating Novel Cations for the Coordination and Extraction of Hg2+ and Cd2+: Synthesis, Characterization, and Extraction Studies, *Environmental Science and Technology*, Vol. 36, Issue 11, 2002, pp. 2523-2529.

2.

Importance and Applications of Some Mediators in Chemistry

Nada F. Atta and Shereen M. Azab

2.1. Mediators and Electrochemistry

Electrochemical sensors and biosensors for pharmaceutical, food, agricultural and environmental analyses have been growing rapidly due to electrochemical behavior of drugs and biomolecules and partly due to advances in electrochemical measuring systems. The merger between fast, sensitive, selective, accurate, miniaturizable and low-cost electrochemistry-based sensing and fields like proteomics, biochemistry, molecular biology, nanotechnology and pharmaceutical analysis leads to the evolution of electrochemical sensors. The chemical modification of inert substrate electrodes with mediators offers significant advantages in the design and development of electrochemical sensors. In operations, the redox active sites shuttle electrons between a solution of the analyte and the substrate electrodes. A further advantage of chemically modified electrodes is that they are less prone to surface fouling and oxide formation compared to inert substrate electrodes [1]. On the other hand, a practical mediator needs to have a low relative molar mass while being reversible, fast reacting, regenerated at low potential, pH independent, stable in both oxidized and reduced forms, unreactive with oxygen and nontoxic. Among the most successful mediators are those based on ferrocene, β-cyclodextrin and phthalocyanine that meet the above criteria.

2.1.1. Chemically Modified Electrodes

A chemically modified electrode (CME) is an electrical conductor (material that has the ability to transfer electricity) has its surface modified for different electrochemical functions. CMEs are modified using advanced approaches to electrode systems by adding a thin film or layer of certain chemicals to change properties of the conductor according to its targeted function [2]. At a modified electrode, an oxidation-reduction substance accomplishes electrocatalysis by transferring electrons from the electrode to a reactant, or a reaction substrate. Modifying electrodes' surfaces has been one of the most active areas of research interest in electrochemistry since 1979, providing control over how electrodes interacts with their environments [2]. Chemically modified electrodes are different from other types of electrodes as they have a molecular monolayer or micrometers thick layers of film made from a certain chemical (depending on the function of the electrode). The thin film is coated on the surface of the electrode. The outcome would be a modified electrode with special new chemical properties in terms of physical, chemical, electrochemical, optical, electrical, transport, and other useful properties. CMEs and electrodes in general heavily depend on electron transport: A general term for electrochemical processes where the charge transports through the chemical films to the electrode. The term coverage is used to express the area-normalized in mol/m^2 of a specific type of chemical site in the thin chemical film in on the surface of the chemically modified electrode.

2.2. Applications of Chemically Modified Electrodes

Advancements in investigations in the field of electrochemical science kept getting more thorough until scientists in the field found no use of bare surfaces to continue their investigations. The reason behind that is that researches that involved electrodes required certain chemical and physical properties that did not naturally exist in the materials used as electrical conductors. To work their way out of the dilemma, they used chemical modification to tailor the materials they used. Atoms, molecules, and nanoparticles are attached to the surface of materials to modify their electronic and structural properties, leading to changing their functionality [2]. In their first stages, CMEs were merely applied in technologies; they were initially made for (tuning surfaces for electrochemical investigations). Also, CMEs provided powerful routes to tune the performance of electrodes.

The modification of electrodes facilitated the following processes in electroanalytical chemistry:

- Providing selectivity of electrodes
- Resisting fouling

- Concentrating species
- Improving electrocatalytic properties
- Limiting access of interferences in complex samples

It also provided a route for other purposes, such as:

- Researching energy conversion
- Researching the phenomena that influence electrochemical processes
- Storing and protecting corrosion
- Developing molecular electronics
- Developing electrochromic devices

The research fields where CMEs are used include the following:

a) Basic electrochemical investigations. Electron transfer occurs between electrodes and electrolytes.

b) Electrostaticity on electrode surfaces. Stationary or slow electric charges are present.

c) Polymer electron transport and ionic transport. Movement of electrons from one species or atom to another, with a special focus on polymers - large molecules with duplicated structural units.

d) Design of electrochemical systems and devices. The creation of systems and devices that use chemically modified electrodes with all the required specifications of the systems or devices.

2.3. Approaches to Chemically Modified Electrodes

There are four ways to chemically modify the surface of electrodes [3]:

Adsorption (Chemisorption):

A method that uses the same kind of valence forces involved in formation of chemical compounds, where the film is strongly adsorbed, or chemisorbed, onto the surface of the electrode, yielding monolayer coverage. This approach involves substrate-coupled self-assembled monolayers (SAMS), where molecules are spontaneously chemisorbed to the surface of the electrode, resulting in a microscopic super lattice structure of layers formed on it.

Covalent bonding:

It is the method that uses chemical agents to create a covalent bond between one or more monomolecular layers of the chemical modifier and the electrode surface. The common agents to use in this method include organosilanes and cyanuric chloride.

Polymer film coating:

It is the method that includes removing chemical species (substrate) from self-assembled monolayers (SAMs) to allow adsorbing molecules on the electrode surface independently of the original substrate structure. The polymer films can be organic, organometallic or inorganic, and it can either contain the chemical modifier or have the chemical added to the polymer in a latter process.

Composite:

A method that has the chemical modifier mixed with an electrode matrix material. An example for this method is having an electron mediator (the chemical modifier) mixed with carbon particles in a carbon paste electrode (the electrode matrix).

2.4. Examples of Some Important Modifiers

2.4.1. Ferrocene

2.4.1.1. Structure

Ferrocene (fc) also called Dicyclopentadienyl iron (Fig. 2.1), is an organometallic with the compound formula $Fe(C_5H_5)_2$. It is the prototypical metallocene, a type of organometallic chemical compound. They are derivatives of transition metals consisting of two cyclopentadienyl organic rings bound on opposite sides of a central metal atom. Such organometallic compounds are also known as sandwich compounds [4]. The rapid growth of organometallic chemistry is often attributed to the excitement arising from the discovery of ferrocene and its many analogues. Ferrocene was first prepared unintentionally. In 1951, Pauson and Kealy at Duquesne University reported the reaction of sodium cyclopentadienide with iron (+2) chloride; ferrocene occurs as highly stable orange crystals with a remarkable stability and a melting point of 174 °C (345 °F). Chemically, ferrocene behaves like benzene and other aromatic compounds in that it undergoes substitution reactions. The removal of one electron from the molecule raises the iron atom to the next-higher oxidation state (i.e., from +2 to +3), leading to the formation of salts containing the blue ferricinium cation, $(C_5H_5)_2Fe^+$. The stability was accorded to the aromatic character of the negatively charged cyclopentadienyls [5]. The structure of ferrocene was confirmed by NMR spectroscopy and X-ray crystallography. Its distinctive "sandwich" structure led to an explosion of interest in compounds of d-block metals with hydrocarbons, and invigorated the development of the flourishing study of organometallic chemistry.

2.4.1.2. Bonding

The carbon-carbon bond distances are 1.40 Å within the five-membered rings, and the Fe-C bond distances are 2.04 Å. Although X-ray crystallography (in the monoclinic space group) points to the Cp rings being in a staggered conformation (Fig. 2.2), it has been shown through gas phase electron diffraction and computational studies that in the gas phase the Cp rings are eclipsed. The staggered conformation is believed to be most stable in the condensed phase due to crystal packing.

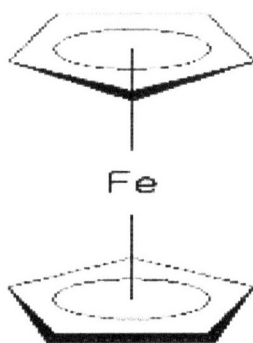

Fig. 2.1. Structure of Ferrocene, reused with permission.

In terms of bonding, the iron center in ferrocene is usually assigned to the +2 oxidation state, consistent with measurements using Mössbauer spectroscopy. Each cyclopentadienyl (Cp) ring is then allocated a single negative charge, bringing the number of π-electrons on each ring to six, and thus making them aromatic. These twelve electrons (six from each ring) are then shared with the metal via covalent bonding. When combined with the six d-electrons on Fe^{2+}, the complex attains an 18-electron configuration. As expected for a symmetric and uncharged species, ferrocene is soluble in normal organic solvents, such as benzene, but is insoluble in water. Ferrocene is an air-stable orange solid that readily sublimes, especially upon heating in a vacuum. It is stable to temperatures as high as 400 °C.

2.4.1.3. Redox Chemistry

Surface modification has been proposed for the production of new superior products in terms of increased corrosion resistance, improvement of optical and electrical/electronic properties, electrocatalysis, sensing, and wettability. The modifications of conducting substrates could be performed by a redox process providing new and special properties to the interfaces. Thus, oxidative and reductive processes have been proposed for the electrochemical grafting of organic molecules onto various electrodes surfaces. Unlike the majority of organic compounds, ferrocene undergoes a one-electron oxidation at a low potential, around 0.5 V vs. a saturated calomel electrode (SCE). It has also been used as standard in electrochemistry as Fc+/Fc = 0.64 V vs. SHE. Some electron-rich organic compounds (e.g., aniline) also are oxidized at low potentials, but only irreversibly. Oxidation of ferrocene gives a stable cation called ferricinium.

On a preparative scale, the oxidation is conveniently effected with $FeCl_3$ to give the blue-colored ion, $[Fe(C_5H_5)_2]^+$, which is often isolated as its PF_6^- salt. Alternatively, silver nitrate may be used as the oxidizer. Ferrocenium salts are sometimes used as oxidizing agents, in part because the product ferrocene is fairly inert and readily separated from ionic products. Substituents on the cyclopentadienyl ligands alters the redox potential in the expected way: electron withdrawing groups such as a carboxylic acid shift the potential in the anodic direction (i.e. made more positive), whereas electron releasing groups such as methyl groups shift the potential in the cathodic direction (more negative).

Thus, decamethyl ferrocene is much more easily oxidized than ferrocene. Ferrocene is often used as an internal standard for calibrating redox potentials in non-aqueous electrochemistry.

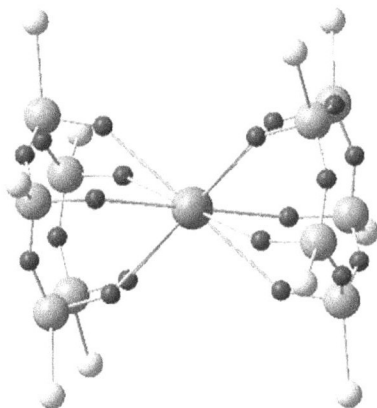

Fig. 2.2. 3D structure of Ferrocene, reused with permission.

2.4.1.4. Applications of Ferrocene and its Derivatives

Ferrocenes have found broad applications as outer sphere redox mediators in solution because of their fast electron transfer rate, their broad range of E8 that is adjusted via the substituents attached to the cyclopentadien ring system and the ease of their synthesis. Attempts to prepare electrode surface-confined ferrocenes are numerous and based on SAM and Langmuir– Blodget techniques or on surface-confined polymers. The latter are, e.g. polysiloxane-ferrocenes, or hydrophobic nafion gels loaded with ferrocenes, or polymers prepared by electropolymerization. Ferrocene and its numerous derivatives have no large-scale applications, but have many niche uses that exploit the unusual structure (ligand scaffolds, pharmaceutical candidates), robustness (anti-knock formulations, precursors to materials), and redox (reagents and redox standards).

2.4.1.4.1. Fuel Additives

Ferrocene and its derivatives are antiknock agents used in the fuel for petrol engines; they are safer than tetraethyl lead, previously used. It is possible to buy at Halfords in the UK, a petrol additive solution which contains ferrocene which can be added to unleaded petrol to enable it to be used in vintage cars which were designed to run on leaded petrol [6]. The iron containing deposits formed from ferrocene can form a conductive coating on the spark plug surfaces.

2.4.1.4.2. Drug Analysis

Some ferrocenium salts exhibit anticancer activity, and an experimental drug has been reported which is a ferrocenyl version of tamoxifen. The idea is that the tamoxifen will bind to the estrogen binding sites, resulting in a cytotoxicity effect. Fc is a well-known electron-transfer mediator in the redox reaction of many electroactive species, such as hydrogen peroxide, dopamine (DA) and ascorbic acid (AA). Recently, ferrocene (Fc) was introduced into the inner cavities of CNTs, including SWNTs (single) and DWNTs (double-walled carbon nanotubes), to obtain a new class of endohedral functionalized CNTs.

2.4.1.4.3. Materials Chemistry

More recently, the so-called layer-by-layer surface modifications known to work with oppositely charged polyelectrolytes have been applied to ferrocenyl organometallic polyelectrolytes. Notably, the layer-by-layer technique allows controlling easily the surface concentration by the number of deposition steps. A drawback of this technique is that the multilayer structure is rather fragile. It is often stable only in an aqueous system and the stability is crucially dependent on the ionic strength and pH. The vinyl ferrocene from ferrocene can be made by a Wittig reaction of the aldehyde, a phosphonium salt and sodium hydroxide [7]. The vinyl ferrocene can be converted into a polymer which can be thought of as a ferrocenyl version of polystyrene (the phenyl groups are replaced with ferrocenyl groups).

2.4.1.4.4. As Ligands

Chiral ferrocenyl phosphines are employed as ligands for transition-metal catalyzed reactions. Some of them have found industrial applications in the synthesis of pharmaceuticals and agrochemicals. For example, the diphosphine 1,1'-bis (diphenylphosphino) ferrocene (dppf) is a valuable ligand for palladium-coupling reactions.

2.4.1.4.5. Voltammetric Measurements

Numerous voltammetric investigations indicate that when the electroactive surface can be oxidized and reduced in a chemically reversible manner, the conductivity of the organic layer switch from an insulating to a conducting state. In such devices, organic surface can truly acts as an ''organic electrode''. The construction of electrodes by incorporating an electroactive substance into a carbon paste matrix was first reported by Kuwana and French in 1964 [8] and has been applied until now for the preparation of chemically modified electrodes for several purposes, such as the determination of trace amounts of some substances, as electrochemical sensors for the analysis of biologically important compounds and for electrocatalysis, etc. A practical mediator needs to be of low relative molar mass, reversible, fast reacting, regenerated at low potential, pH independent, stable in both oxidized and reduced forms, unreactive with oxygen and non-toxic. Among the most successful mediators are those based on ferrocene and its derivatives, all of which meet the above criteria.

Electrochemical oxidation of ferrocene (Fc) at platinum and glassy carbon electrodes in ionic liquid has been studied by cyclic voltammetry (CV), convolutive potential sweep voltammetry (CPSV), chrono-potentiometry (CP) and chronoamperometry (CA). A carbon paste electrode spiked with ferrocene carboxylic acid (FCAMCPE) was constructed by incorporation of ferrocene carboxylic acid can catalyze the oxidation of ascorbic acid in aqueous buffered solution. Also a new ferrocene-derivative compound, 1-(4-bromobenzyl)-4-ferrocenyl-1H-[1,2,3]-triazole (1,4-BBFT),was synthesized and used to construct a modified-graphene paste electrode for simultaneous determination of isoproterenol, acetaminophen and theophylline, while dopamine was investigated at electrodes modified with ferrocene-filled double-walled carbon nanotubes. Redox-active materials in enzyme biosensors commonly use Fc derivatives, which mediate electron transfer between the electrode and enzyme active site. Either voltammetric or amperometric signals originating from redox reactions of Fc are detected or modulated by the binding of analytes on the electrode. Fc-modified thin films have been prepared by a variety of protocols, including in situ polymerization, layer-by-layer (LbL) deposition, host-guest complexation and molecular recognitions. In situ polymerization provides a facile way to form Fc thin films, because the Fc polymers are directly deposited onto the electrode surface. LbL deposition, which can modulate the film thickness and Fc content, is suitable for preparing well-organized thin films. Other techniques, such as host-guest complexation and protein-based molecular recognition, are useful for preparing Fc thin films. Fc-modified Au nanoparticles have been widely used as redox-active materials to fabricate electrochemical biosensors. Fc derivatives are often attached to Au nanoparticles through a thiol-Au linkage. Nanoparticles consisting of inorganic porous materials, such as zeolites and iron oxide, and nanoparticle-based composite materials have also been used to prepare Fc-modified nanoparticles.

Atta et al presented an electrochemical sensor based on the electrodeposition of ferrocene over gold nanoparticles modified carbon paste electrode. The proposed sensor was utilized for the simultaneous determination of morphine, ascorbic acid and uric acid [9]. In addition, Atta et al constructed a sandwich sensor for dopamine based on the insertion of a film of ferrocene carboxylic acid (FC1), ferrocene (FC2) or cobaltocene (CC) between two layers of conducting polymer of poly(3,4-ethylenedioxythiophene) in the presence of surface active agent (Fig. 2.3). FC1 mediator exhibited the highest electrocatalytic activity as a result of the presence of the ferrocenium ion and the polar substituted COOH group in the matrix resulting in enhanced electronic conduction [10]. Furthermore, Atta et al fabricated an electrochemical sensor for morphine based on the inclusion of ferrocene carboxylic acid between two layers of poly(3,4-ethylenedioxythiophene) and gold nanoparticles (Fig. 2.3) [11].

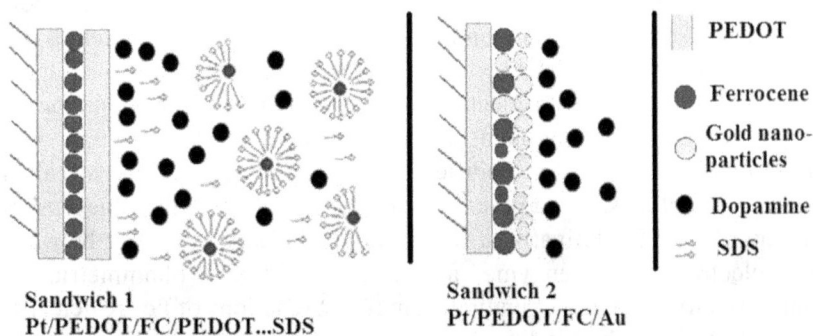

Sandwich 1
Pt/PEDOT/FC/PEDOT...SDS

Sandwich 2
Pt/PEDOT/FC/Au

PEDOT

Ferrocene

Gold nano-particles

Dopamine

SDS

Fig. 2.3. Sandwich electrode based on polymer/ferrocene/polymer and polymer/ferrocene/gold for dopamine sensing [10, 11], reused with permission.

2.4.1.4.6. As DNA Sensor

Electrochemistry allows the formation of radical species, in the vicinity of the electrode, from various precursors by electron transfer coupled to bond dissociation process leading to the attachment of thin organic film, in the range of few nanometers, onto electrode surface. Ferrocene-dT

(Fc-dT), which is a base nucleotide modified by a redox-active Fc moiety, is used as an electrochemical probe for nucleic acid analysis by incorporating it into an oligonucleotide. Many research groups have developed electrochemical DNA sensors with Fc-labeled oligo-nucleotides. The target binds specifically with the surface-confined and redox-labeled aptamers in these sensors, resulting in a conformational change of the aptamer probes. The conformational change leads to a change in the distance between the labeled redox moieties and the electrode [12]. This facilitates effective electron transfer and enhancement of the electrochemical signals.

2.4.1.4.7. Ion-Selective Electrodes

Polymeric ion selective membranes play an important role in fields such as clinical diagnostics and environmental monitoring for the detection of small and hydrophilic ions. While traditional membrane electrodes are backside contacted with an aqueous electrolyte, recent efforts aimed at replacing the inner solution by a solid contact ion-to-electron transducer, which includes conducting polymers and nanostructured materials as potential candidates to replace the inner filling solution and the Ag/AgCl element. Because ion-selective membranes have shown attractive analytical characteristics in open circuit potentiometry as well as in dynamic electrochemistry, the application of solid contact ion-to electron transducers may allow one to achieve a new generation of ion sensors, as achieved earlier with potentiometric ion-selective electrodes. Langmaier et al. used freely dissolved dimethyl ferrocene, DMFc, a well-established electroactive material that exhibits excellent redox capacity. DMFc was embedded in a polymeric non-permselective film containing PVC, an apolar plasticizer (DOS) and a lipophilic salt. The oxidation of DMFc at the electrode creates anion-exchanging groups at the metal-membrane interface during the amperometric pulse. To fulfill electroneutrality, anions are concurrently extracted from the sample into the film in order to compensate for the unbalanced charge. This methodology was successfully used to determine heparin (a polyanion with about 70 charges per molecule) in the physiological concentration range (1–10 U mL^{-1}). Recently, Pawlak et al. introduced ferrocene groups covalently attached to the PVC backbone by "click chemistry" (Huisgen cycloaddition) [13]. This prevents the possible loss of the ferrocene derivative by leaching into the sample solution and provides a materials approach to confine the ferrocene functionality to a layer close to the contacting metal electrode at the inner side of the sensing film.

2.4.2. Cyclodextrins

Cyclodextrins, as they are known today, were called "cellulosine" when first described by A. Villiers in 1891. Soon after, F. Schardinger identified the three naturally occurring cyclodextrins -α, -β, and -γ. These compounds were therefore referred to as "Schardinger sugars". For 25 years, between 1911 and 1935, Pringsheim in Germany was the leading researcher in this area, demonstrating that cyclodextrins formed stable aqueous complexes with many other chemicals. By the mid-1970s, each of the natural cyclodextrins had been structurally and chemically characterized and many more complexes had been studied. Since the 1970s, extensive work has been conducted by Szejtli and others exploring encapsulation by cyclodextrins and their derivatives for industrial and pharmacologic applications. Recently, the largest well-characterized cyclodextrin contains 32 1,4-anhydro-glucopyranoside units, while as a poorly characterized mixture, at least 150-membered cyclic oligosaccharides are also known.

2.4.2.1. Structure

Cyclodextrins are a family of cyclic oligosaccharides typically containing six (α-cyclodextrin), seven (β-cyclodextrin), or eight (γ-cyclodextrin) 1,4-linked D-glucopyranose subunits [14] obtained by degradation of starch by the enzyme cyclodextrin gluco-syltransferase (Fig. 2.4). They are classified into hydrophilic, hydrophobic and ionic derivatives. Because the glucose units adopt the chair conformation, the cyclodextrins are shaped like a hollow truncated cone with a hydrophilic outer surface, which makes them water-soluble. The flexible 6-OH hydroxyl groups are al so capable of forming linking hydrogen bonds around the bottom rim but these are destabilized by dipolar effects, easily dissociated in aqueous solution and not normally found in cyclodextrin crystal s. The hydrogen bonding is all 3-OH (donor) and 2-OH (acceptor) in α-cyclodextrin but flips between this and all 3-OH (acceptor) and 2-OH (donor) in β- and γ -cylodextrins. The central cavity of the cone is lined by the skeletal carbon atoms and ethereal oxygen atoms of the glucose residues, which gives it a lipophilic character. This combination of a hydrophilic exterior with a hydrophobic interior enables cyclodextrins to form inclusion complexes with hydrophobic guest molecules. As a result, the physical and chemical properties of the guest molecule can be greatly modified, mostly in terms of water solubility. This is the primary reason why cyclodextrins have attracted great interest in a variety of industries, including those related to food, pharmaceuticals, cosmetics, chemicals, and agriculture. Cyclodextrins are produced from starch or starch derivatives by means of an enzymatic conversion catalyzed by cyclodextrin glycosyltransferase. The principal advantages of natural/ parent cyclodextrins (α, β and γ) are their low toxicity, low pharmacological activity and well-defined chemical structure.

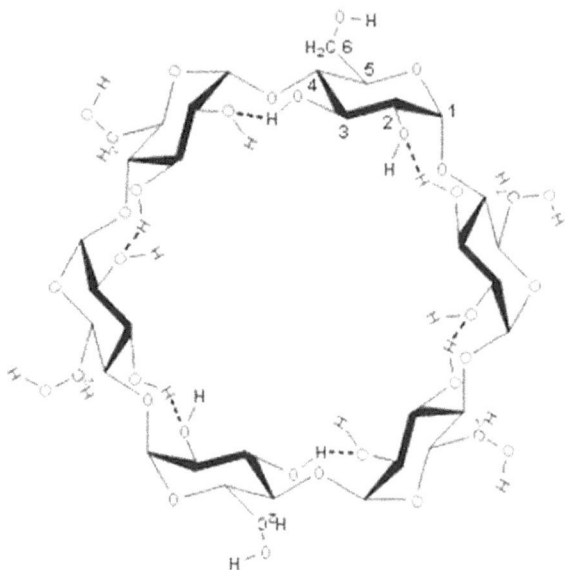

Fig. 2.4. Chemical structure of cyclodextrin, reused with permission.

Cyclodextrins are toroidal molecules with a truncated cone structure having lipophilic inner cavities and hydrophilic outer surface (Fig. 2.4). Their cylindrical structures with cavities of about 0.7 nm deep and 0.5–0.8 nm inside diameter yield various unique properties.

2.4.2.2. Types of Cyclodextrins

Typical cyclodextrins contain a number of glucose monomers ranging from six to eight units in a ring, creating a cone shape (Fig. 2.5):

- α (alpha)-cyclodextrin: 6-membered sugar ring molecule
- β (beta)-cyclodextrin: 7-membered sugar ring molecule
- γ (gamma)-cyclodextrin: 8-membered sugar ring molecule

α- and γ-cyclodextrin are being used in the food industry. As α-cyclodextrin is a soluble dietary fiber, it can be found as Alpha Cyclodextrin (soluble fiber) on the list of ingredients of commercial products.

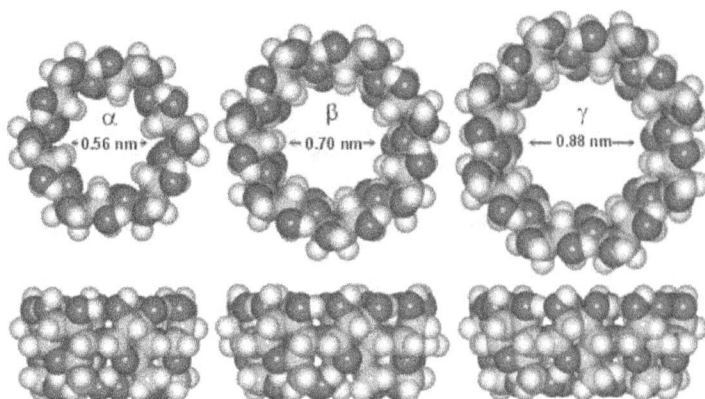

Fig. 2.5. TS and LS of the 3 types of cyclodextrin and their diameters, reused with permission.

2.4.2.2.1. Alpha-cyclodextrin

α-Cyclodextrin (α-CD) (dietary fiber) is a polysaccharide of six glucose units that are covalently attached end to end via α-1, 4 linkages. In water, these fibers take on a toroid or truncated cone configuration. In aqueous medium, the exterior surface of cyclodextrins is hydrophilic while the interior core is hydrophobic. α-cyclodextrin was very effective at solubilizing the free fatty acids that are generated in the colorimetric determination of triglycerides. According to these authors, the number of cyclodextrin molecules per fatty acid appears to be dependent upon the length of the fatty acids that are involved. Due to the small pore size it was generally believed that α-cyclodextrin could not form a complex with triglyceride.

α-Cyclodextrin is a multifunctional, soluble dietary fiber marketed for use as a fiber ingredient, an odor or flavor masking agent as well as for emulsification applications. It is registered as a dietary fiber in the European Union since 2008. Also α-Cyclodextrin is marketed for a range of medical, healthcare and food and beverage applications which rely on its ability to bind to fats and reduce their bioavailability both in the body and in food and beverage products. Although many dietary fibers appear to bind fat on a 1:1 ratio α-cyclodextrin is the only one that binds on a 1:9 (fiber:fat) ratio. This high ratio of binding makes α-cyclodextrin practical as a weight loss supplement.

Due to its surface active properties, α-cyclodextrin can also be used as an emulsifying fiber, for example in mayonnaise. It stabilizes oil-in-water emulsions very efficiently. The three-dimensional, donut-shaped cyclodextrins have a hydrophobic cavity inside and a hydrophilic cover on the outside. A fatty acid tail of triglycerides is attracted by the cavity and encapsulated there. This leads to the build-up of a surfactant-like structure which has emulsion-stabilizing properties. Depending on the oil-to-water ratio and the amount of α-cyclodextrin used, the viscosity, and therefore the organoleptic properties, of the emulsion are altered. From ketchup-like viscosity to icing-like viscosity, all grades can be adjusted. Often with significant less fat content and thus reduced calories. Furthermore, stable emulsions are feasible even at elevated temperatures.

α-Cyclodextrin can also be used as whipping fiber, for example in desserts and confectionary applications. Tests showed that alpha cyclodextrins can cause a volume effect in various different food compositions with or without fat and at a very broad pH range. This can be used for fat free or fat containing dessert compositions and for the reduction or the replacement of egg white in confectionary and bakery applications.

2.4.2.3. Gamma-cyclodextrin

Gamma-cyclodextrin (γ-CD) is a cyclic alpha-(1,4)-linked oligo-saccharide consisting of eight glucose molecules. Like other cyclodextrins, γ -CD can form inclusion complexes with a variety of organic molecules because the inner side of the torus-like molecule has less polarity than the outer side. In foods, γ -CD may be used as a carrier for flavors, vitamins, polyunsaturated fatty acids, and other ingredients. It also has useful properties as a stabilizer in different food systems.

An interaction of ingested gamma-CD with the absorption of fat-soluble vitamins or other lipophilic nutrients is not to be expected because the formation of inclusion complexes is a reversible process, γ -CD is readily digested in the small intestine, and studies with beta-CD, a nondigestible cyclodextrin, have shown that the bioavailability of vitamins (A, D, and E) is not impaired. On basis of these studies it is concluded that γ -CD is generally recognized as safe (GRAS) for its intended uses in food.

2.4.2.4. Beta-cyclodextrin

The attractive property of CDs is the presence of many hydroxyl groups on glucose units. Modification of hydroxyl groups is expected to affect the capability of molecular recognition. Methylation or hydroxyalkylation of hydroxyl groups improve the solubility and stability of inclusion complexes with guest molecules [15].

The secondary hydroxyl groups are on the wider side of the ring, whereas the primary hydroxyl groups are on the opposite narrower side of the torus. This low polarity central void is able to encapsulate (partially or wholly) a wide variety of guest molecules of suitable size, shape, structure and dimensions, resulting in a stable association without the formation of covalent bond. The hydroxyl groups allow direct substitution reactions and/or chemical modifications to different positions yielding various polymerized derivatives for specific domains of application. The interesting characteristics can enable them to form stable host–guest inclusion complexes or nanostructured supramolecular assemblies in their hydrophobic cavity. Therefore, it can show high molecular selectivity and enantioselectivity and improve the solubility and stability of functional materials. The inclusion complex (Fig. 2.6) is present in solution in dynamic equilibrium with its constituents and is characterized by the absence of covalent bonds and by a well-defined host: guest stoichiometry.

Several types of forces are involved in the inclusion complex formation, and their relative contribution depends on the guest and CD type. These forces include hydrophobic interactions, reduction of conformational strain, hydrogen bonding, dipole–dipole and electrostatic interactions, van der Waals and dispersion forces; however, the main driving force is considered the replacement of the unfavoured polar–apolar interactions between both the included water molecules and the CD cavity on one hand, and water and the hydrophobic guest on the other one, by more favored apolar–apolar interactions between the guest and the cavity [16]. β-Cyclodextrin (β-CD) is most widely used because it is readily available and has suitable cavity size. Moreover, it provides pharmaceutically useful complexation characteristics for the widest range of drugs. However, the market share of α-cyclodextrin is currently much smaller than that of β-cyclodextrin due to its low production yield and high price. β-CD, have limited aqueous solubility, attributed to the relatively strong intramolecular hydrogen bonds between the secondary hydroxyl groups, that diminish their ability to form hydrogen bonds with the surrounding water molecules. Therefore, several chemically modified CD derivatives have been developed with a view to improve the physicochemical properties of parent CDs. Partial random substitution of CD hydroxyl groups, even with hydrophobic moieties (i.e. methoxy functions) results in dramatic increase of CD solubility, by transforming crystalline CDs into amorphous mixtures of isomeric derivatives, and preventing formation of intramolecular hydrogen bonds, thus making the residual free hydroxyl groups available to interact with water. Moreover, the presence of substituents can extend the CD hydrophobic cavity, improving their complexing ability.

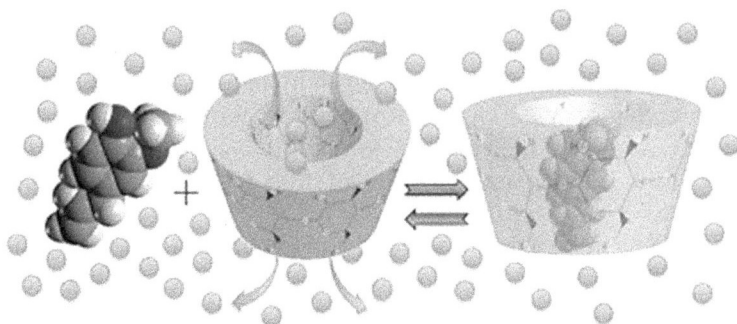

Fig. 2.6. Schematic representation of the formation of a typical cyclodextrin inclusion complex, reused with permission.

2.4.2.5. Applications

2.4.2.5.1. Cholesterol Free Products

In the food industry, cyclodextrins are employed for the preparation of cholesterol free products: the bulky and hydrophobic cholesterol molecule is easily lodged inside cyclodextrin rings that are then removed. Both β-cyclodextrin and methyl-β-cyclodextrin (MβCD) remove cholesterol from cultured cells. The methylated form MβCD was found to be more efficient than β-cyclodextrin. The water-soluble MβCD is known to form soluble inclusion complexes with cholesterol, thereby enhancing its solubility in aqueous solution. MβCD is employed for the preparation of cholesterol-free products: the bulky and hydrophobic cholesterol molecule is easily lodged inside cyclodextrin rings that are then removed. MβCD is also employed in research to disrupt lipid rafts by removing cholesterol from membranes [17].

2.4.2.5.2. Pharmaceutical Field

The always growing interest toward CDs, particularly in pharmaceutical field, is directly related to the number of potential advantages which can be obtained by the improvement of unfavorable physical chemical properties of several drug molecules through CD inclusion complexation, together with the increased availability of CDs at lower cost. Spectroscopic techniques, and particularly NMR, have played and still play a fundamental role in the characterization of inclusion complexes in solution. However, other techniques such as electroanalytical techniques, HPLC and CE and isothermal calorimetric techniques, especially when used in combination with spectroscopic techniques, can offer an important supporting tool for allowing a more complete extraction of valuable information concerning host–guest interactions in solution and formation of drug–CD inclusion complexes.

The bio-adaptability and versatility of cyclodextrins makes them capable of alleviating the undesirable properties of drug molecules in various areas of drug delivery through the formation of inclusion complexes. In fact, numerous derivatives of CDs are examined continuously to improve physicochemical properties of drug in order to obtain both higher solubility and stability. Cyclodextrin-based carriers can enhance the capability of encapsulating guest molecule, improve the stability of drug and efficiently regulate the drug release rate [18]. Cyclodextrins enhance the bioavailability of insoluble drugs by increasing the solubility and permeability. They increase the permeability of hydrophobic drug by making them available at the surface of biological barrier (e.g. skin and mucosa) from where it partitions into the membrane without disrupting the lipid layers of the barrier. In case of water-soluble drugs, cyclodextrins increase the permeability by direct action on mucosal membrane and enhance the drug absorption and bioavailability.

Moreover CD complexation can also be exploited to mask unpleasant tastes or smell, reduce evaporation and stabilize volatile substances, protect molecules sensitive to light or oxygen, convert liquid substances and oils in free-flowing powders, reduce gastric, dermal or ocular irritation and prevent incompatibilities and interactions between substances. The numerous potential advantages related to their use and the increased availability of CDs at lower cost play a decisive role in the growing interest toward these molecules, particularly in the pharmaceutical field. In fact, to fully exploit the potential of CD inclusion complexes, it is important to have at disposal adequate analytical techniques for their suitable and careful characterization. In particular, the determination of the stability constants of the inclusion complexes is a crucial point for the evaluation of their effectiveness, since the different possible effects related to the complex formation all rely on the stability of the complexes formed. β-cyclodextrins can solubilize hydrophobic drugs in pharmaceutical applications, and crosslink to form polymers used for drug delivery [19].

Due to the unique structure of molecules combined with their poly-functionality, β-CD molecules have the ability to form cross-linked networks. It can be cross-linked by direct reaction between its hydroxyl groups with a coupling agent to form water-soluble/insoluble polymeric structures [20]. Polymerized cyclodextrins can be cationic, anionic or non-ionic in nature. Due to the presence of charge, these charged polymerized cyclodextrins possess special complexing and solubilizing ability. The ionic interactions outside the cavities and the hydrophobic interactions inside the cavities, contributes to the drug solubilization. They may form stronger complex with oppositely charged molecules and weaker complex with molecules of same charge.

Galal et al fabricated an electrochemical sensor of dopamine [21] and paracetamol [22] in the presence of common interferents based on the modification of poly(3,4-ethylenedioxythiophene) modified gold electrode with nafion film and β-cyclodextrin (CD). Synergistic effect was achieved through the high conductivity of polymer film and nafion as well as the preconcentration effect of cyclodextrin. CD forms a host-guest inclusion complex through the formation of H-bonding with the compounds of study. Fig. 2.7 shows the inclusion complex between CD and dopamine, uric acid and ascorbic acid [21] or paracetamol, dopamine and ascorbic acid [22] through the formation of H-bonds. The conducting polymer facilitates the electron mediation through its electron cloud. Nafion film enhances the electrical conductivity and facilitates the formation of CD film on its surface as a result of its film forming ability and good adhesion over the polymer film. A supramolecular host-guest complex was formed between the studied analytes and CD through hydrogen bonding and electrostatic and inclusion interactions. A selective interaction was achieved through this complex formation as the studied analyte penetrates the less polar CD cavity resulting in enhanced current response on paracetamol at the CD modified electrode [21, 22]. Moreover, Galal et al modified the graphene sheets which have high surface area with the high ionic conductive ionic liquid crystal (ILC) and CD with its preconcentrating effect (Fig. 2.8) to be used as an electrochemical sensor for neurotransmitters like (dopamine, epinephrine, norepinephrine, l-dopa, serotonin and 3,4-dihydroxy-phenyl acetic acid) [23] and a muscle relaxant drug "methocarbamol" [24]. CD forms a highly stably host-guest inclusion complex with neurotransmitters and methocarbamol resulting in enhanced electron transfer kinetics [23, 24].

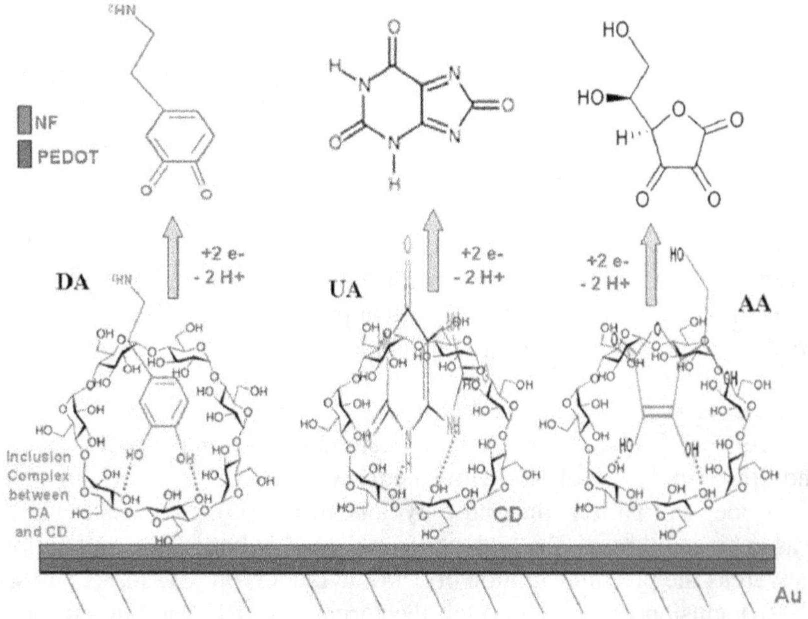

Fig. 2.7. Schematic representation of the inclusion complex formed between CD and DA, with AA and UA as guest molecules [21], reused with permission.

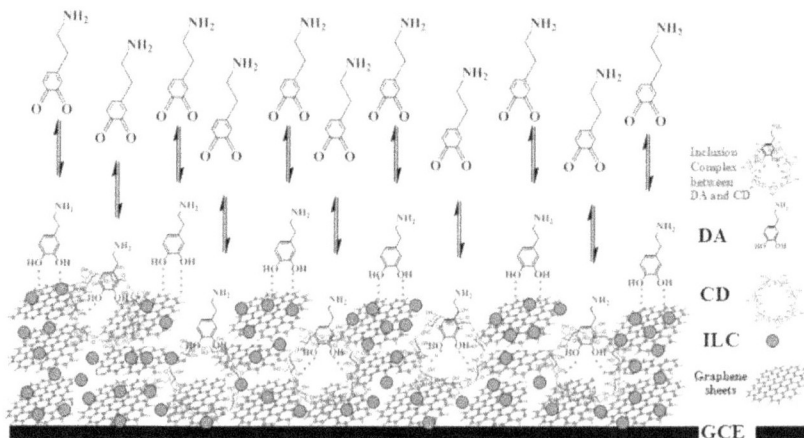

Fig. 2.8. Schematic diagram of graphene/ILC/CD modified electrode [23], reused with permission.

2.4.2.5.3. Environmental Protection

β cyclodextrins can also be employed in environmental protection: these molecules can effectively immobilize inside their rings toxic compounds, like trichloroethane or heavy metals, or can form complexes with stable substances, like trichlorfon (an organophosphorus insecticide) or sewage sludge, enhancing their decomposition.

2.4.2.5.4. Supramolecular Chemistry

Cyclodextrins also represent a paradigmatic example of carbohydrate derivatives, exhibiting a close relationship between molecular status and supramolecular functional properties. Thus, considered as a class of nano-biomaterial, susceptible of further manipulation to modulate their topology and recognition features with the environment. Coating inside a nanocavity with sugars or building up a glyconanocavity from carbohydrate building blocks is, however, much less evident. This is the reason that cyclodextrins are the only representatives of nanomaterials that allow both fundamental studies and commercial applications. This ability of forming complexes with hydrophobic molecules has led to their usage in supramolecular chemistry. The application of cyclodextrin as supramolecular carrier is also possible in organometallic reactions. The mechanism of action probably takes place in the interfacial region [25].

2.4.2.5.5. HPLC Columns

β-cyclodextrins are used to produce HPLC columns allowing chiral enantiomers separation, and are also the main ingredient in P&G's product Febreze which claims that the β-cyclodextrins "trap" odor causing compounds, thereby reducing the odor [26].

2.4.2.6. Analytical Techniques to Characterize Drug–CD Complexes

Analytical characterization of host–guest interactions is of fundamental importance for fully exploiting the potential benefits of complexation, helping in selection of the most appropriate cyclodextrin. The assessment of the formation of a drug–CD inclusion complex and its full characterization is not a simple task and often requires the use of different analytical methods, whose results have to be combined and examined together, since each method explores a particular feature of the inclusion complex. The concomitant use of different techniques can allow a better and more in-depth understanding of host–guest interactions and help in selection of the most appropriate CD for a given guest molecule. The different available methods are generally based on the detection of the variation in any suitable physical or chemical property of the guest as a consequence of the inclusion complex formation. Obviously, it is essential that the observed variation is large enough to be detected or estimated with sufficient precision. The main analytical techniques used for characterization of CD inclusion complexation in solution include spectroscopic, electroanalytical, separation, polarimetry and isothermal titration calorimetry techniques.

2.4.2.6.1. Spectroscopic Techniques

All the spectroscopic methods are based on the measurement of a variation, upon inclusion complex formation, in a given property of the system. Therefore, for their applicability, it is necessary that this variation can be detectable with sufficient precision.

2.4.2.6.1.1. Ultraviolet/visible Spectroscopy

UV–vis spectroscopy is a simple, economic, fast and useful method of studying the formation of host–guest complexes in solution, when the complexation gives rise to a significant modification of the absorption spectrum of the guest molecule. In fact, depending on the position of the drug chromophore, the transfer of the guest molecule from an aqueous medium to the non-polar CD cavity can modify its original UV absorption spectrum, due to partial or total replacement of the salvation shell of the molecule by the CD molecule, which leads to new solute environment interactions. Modifications of the UV spectrum of a drug in presence of CDs can therefore provide evidence of the formation of an inclusion complex. Bathochromic shifts due to inclusion complex formation with CDs have been reported for several kinds of host molecules, such as, for example, naproxen, itraconazole, rutin, genistein, erythromycin. On the other hand, shifts to shorter wavelength of the absorption maximum have been observed during complexation with CDs of salicylic acid, nicotinic acid, indomethacin, hydrocortisone. Increase in intensity of the absorption maximum upon CD complexation, with or without changes in its max, have been described for numerous drug molecules.

2.4.2.6.1.2. Circular Dichroism Spectroscopy

Circular dichroism can represent a powerful technique to prove the CD inclusion complexation of both chiral and non-chiral guest molecules and obtain information about the structure of the complex in aqueous solution. In the case of chiral guest molecules, changes in their circular dichroism spectra may be detected, attributed to the increased optical activity induced by the formation of inclusion complexes with CDs. The effect is only observed when the chromophore moiety of the guest molecule is actually included in the CD cavity. Circular dichroism spectra of complexes with CDs are characterized by their sign, magnitude and wavelength of the maximum. The magnitude of the effect seems to be related to the rigidity of the complex formed.

2.4.2.6.1.3. Fluorescence Spectroscopy

Fluorescence spectroscopy is useful for investigating the formation in solution of CD inclusion complexes of fluorescent guests. In fact, an enhancement in fluorescence is generally observed upon inclusion of a fluorescent guest molecule into the CD cavity, due to shielding from quenching and non-radioactive decay processes [27]. The fluorimetric method benefits of a high sensitivity, but its application field is limited to fluorescent molecules. Moreover, the preparation of samples for fluorimetry is tedious and time-consuming, because a very strict care is required to avoid spurious interferences. Enhancement in fluorescence upon complexation with βCD or its hydroxypropylated or methylated derivatives have been reported for example for naproxen and benzocaine. The formation of inclusion complexes of warfarin with of α-, β-, and γ-CDs resulted in enhancement of the fluorescence of the drug, offering also a tool for the development of bioanalytical methodologies for its quantification in biological liquids.

2.4.2.6.1.4. Nuclear Magnetic Resonance (NMR) Spectroscopy

NMR spectroscopy is considered one of the most useful and complete analytical techniques to investigate interactions between CDs and guest compounds. No other spectroscopic technique can provide the same wealth of information on the supramolecular systems. In fact, NMR spectra allow elucidation of the structure of the complex in solution, providing specific information on the orientation of the guest molecule inside the cavity, while the other spectroscopic techniques can give only indirect information on the molecular structure of the inclusion complexes.

The simplest NMR experiment to fast obtain direct evidence of the inclusion of a guest into the CD cavity is the observation of the difference in the proton (1H NMR) or carbon (13 C NMR) chemical shifts between the free guest and host species and the presumed complex. Analysis of chemical shift changes of both host and guest molecules can not only give evidence of the complex formation but also supply useful information about the stoichiometry, stability, mechanism and geometry of the complex. Measurements of chemical shift changes of the guest as a function of increasing CD concentration (NMR titrations) allow the evaluation of the complex association constant, providing at the same time insight into the stoichiometry and conformation of the formed complex [28]. The main drawback of this technique is the poor solubility of the samples in deuterated water (1H NMR) or in water (13C NMR), requiring often the use of other solvents, which could modify the host–guest interactions with respect to the simple aqueous medium. 13 C NMR experiments, performed according to the mole ratio and the continuous variation methods to study the interactions in solution of naproxen with natural (α-, β-, and γ-) and derivative CDs, provided important details on the stoichiometry, mechanism, and geometry of the respective complexes; nuclear relaxation measurements enabled evaluation of the molecular dynamics of the drug in the complexes, confirming its embedding within the hydrophobic core of the macrocycle. 13 C NMR and 1H NMR studies provided clear evidence of partial inclusion of ascorbic acid into the hydroxypropyl-βCD cavity, and of formation of ternary complexes with triethanolamine, which markedly increased the drug chemical stability [29].

2.4.2.6.1.5. Electron Spin Resonance (ESR)

Electron spin resonance (ESR) or electron paramagnetic resonance (EPR) is a spectroscopic technique used for characterizing chemicals species containing unpaired electrons, including free radicals. Therefore, it can be a useful and very specific method to investigate inclusion complexation of CDs with radical species in aqueous solutions. Since the hyperfine coupling constant of radicals is very sensitive to the medium polarity, its alterations due to its movement toward an environment less polar than water, such as the CD cavity, is indicative of the inclusion complex formation.

2.4.2.6.2. Electroanalytical Techniques

CDs have been the subject of wide electroanalytical studies, regarding their behavior in homogeneous solutions as well as in thin films on the electrode surfaces. Electroanalytical methods are considered very effective for the determination of the stability constants of inclusion complexes of CDs in solution, especially when the guest molecule is electroactive. Among such methods, polarographic and voltammetric ones are the most largely used.

2.4.2.6.2.1. Polarography and Voltammetry

Polarography and voltammetry are techniques suitable for studying the inclusion complexation of CDs with electroactive guest molecules and may be a powerful tool for the elucidation of the nature of the inclusion complexes. Moreover, being highly sensitive and poorly material-consuming methods, they are particularly useful to evaluate the association constants of inclusion complexes at very low concentration levels of the electroactive guest. In polarography, complexation with CDs gives rise to a reduction of the amount of conducted current upon reduction of the guest molecule, as a result of its decreased diffusion rate when complexed with CD. Moreover, changes in the half-wave potential of the guest molecule may be observed as a consequence of the electron redistribution occurring in the presence of CDs, due to the formation of inclusion complexes, and the intensity of such changes reflects the tendency of the guest molecules to complex with the CD. Analogously, in voltammetry, shifts in a negative direction of the cathodic peak potential of the guest molecule and, at the same time, decrease in intensity of the peak current are observed as a consequence of the inclusion complex formation. Differential pulse cathodic stripping voltammetry and cyclic voltammetry have been applied, for example, for the determination of the stability constants and the Gibbs energy of the inclusion complexes of lumazin with αCD or βCD [30].

2.4.2.6.2.2. Potentiometry

Potentiometric measurements can be a very useful approach to determine the association constants of CD complexes with acidic or basic drug molecules, by monitoring the pH variation as function of increasing CD concentrations, keeping constant the guest concentration. The formation and stability of CD complexes with basic or acid drugs is strongly influenced by the pH of the complexation medium, which will determine the ratio between the ionized and unionized forms present in solution. Both these forms can be included into the CD cavity, and therefore different equilibrium complexes will be present together in solution, and they will have different stability constants, related to the different hydrophobicity of ionized and unionized forms.

The values of the association constants of the equimolar inclusion complexes formed by hydroxypropyl-βCD and the ionized and unionized forms of the acid drugs flurbiprofen and ibuprofen have been determined at different temperatures by pH potentiometric measurements; at all temperatures the complexes with the carboxylic species showed higher stability constants than those of their ionized counterparts, in virtue of their higher hydrophobicity and then higher affinity for the hydrophobic CD cavity [31].

2.4.2.6.2.3. Electrical Conductivity

Electrical conductivity measurements have been employed to determine equilibrium constants of complexes between CDs and a variety of ionic surfactants and other amphiphilic molecules. The binding constants of the complexes of βCD or hydroxypropyl-βCD with dodecyltrimethyl ammonium bromide in aqueous solution were evaluated by electrical conductivity, considering the association of the surfactant counterion with the inclusion complex and the variation of the ionic molar conductivities with the concentration. Conductivity measurements have been performed to investigate the behavior of three tricyclic antidepressant drugs (imipramine, desipramine, and amitriptyline hydrochlorides) in aqueous solutions in absence and presence of βCD; all the drugs exhibited aggregation phenomena and the effect of βCD on drug aggregation has been evaluated by determining the apparent critical aggregation concentration and the dissociation degree of the ternary βCD/drug/H$_2$O system [32].

2.4.2.6.3. Separation Techniques

CDs are able to form highly selective inclusion complexes with a variety of guest molecules, and, in particular, are endowed with a discriminating power toward enantiomeric substances, in virtue of the inherent chirality of their structure. These properties led to their wide use as stationary phases and as mobile phase modifiers in chromatography, particularly in high performance liquid chromatography (HPLC) and in capillary electrophoresis (CE), for the separation of drug enantiomers or of closely related compounds, including geometrical structural isomers. Moreover, HPLC and CE revealed to be useful methods for investigating molecular interactions such as complexation of guest compounds with CDs, and they can provide a valuable way to determine accurate and reproducible apparent association constants of the inclusion complexes.

2.4.2.6.3.1. High Performance Liquid Chromatography (HPLC)

HPLC is a powerful tool to study the interactions of CDs and CD complexes with stationary phases, as well as to determine the stoichiometry and association constants of CD complexes in solutions. In HPLC experiments, CDs can be in the stationary phase, chemically bound to silica gel, or, more frequently, they can be added to the mobile phase: as a result of host–guest interactions, the retention time of the guest will change, becoming longer or shorter, depending on complexation occurring in the stationary or in the mobile phase, respectively (Fujimura method) [33]. Another approach, known as Hummel–Dreyer method [34], uses a column equilibrated with an eluent carrying the guest molecule; when the complexing agent is added in the eluent solution, the chromatogram will show a positive peak, due to the complex, and a negative peak at the retention time of the free molecule, corresponding to the amount of complexed guest. The stability constant is calculated by the variation of intensity of the negative (or positive) peak as the concentration of the ligand is increased.

2.4.2.6.3.2. Capillary Electrophoresis (CE)

CE can be defined as a technique for the high-efficient separation of molecules in capillary tubes containing an electrolyte solution under the influence of an electric field. CE offers significant advantages compared to other separation methods, such as high efficiency, allowing the observation of very fine enantioselective effects, high flexibility, high speed and miniaturization. The separation is based on the difference in mobility of ionic species or in the affinity of charged or uncharged molecules toward charged electrolytes. Analyses of intermolecular interactions based on affinity effects such as electrostatic interactions, Van der Waals forces, hydrogen bonding, are more specifically designated as affinity capillary electrophoresis (ACE). CDs played an important role in the development of various CE analytical methods. CE proved to be a versatile and useful analytical tool for the analysis of CDs and their inclusion complexes with charged guest molecules, with the advantage of requiring only minute amounts of sample. CE has been employed for the estimation of the association constants of CDs complexes with different guest molecules, according to two different approaches. The stability of the inclusion complexes of βCD with benzoate and a set of structurally related hydroxybenzoates has been investigated for the determination of host–guest association constants; the study allowed for a direct comparison of the data, based on the structural differences between the guest molecules.

2.4.2.6.4. Polarimetry

Polarimetry has not particularly attracted the attention of researchers as a tool for obtaining thermodynamic and structural information and/or determine the binding constants of CD complexes. For example, a polarimetric study was conducted as a supporting tool, together with other analytical techniques, for a more complete characterization of the complex formation between βCD and celecoxib [35]. On the contrary, Lo Meo et al recently proved that polarimetry is a very simple technique, able to provide an easy, fast, accurate and low-consuming materials method for the evaluation of the stability constants of CD inclusion complexes. The proposed approach is based on the observation that the optical activity of CDs, due to the presence of chiral glucose units in their structure, is strongly affected by their mutual spatial arrangement. In fact, the molar optical activities of native CDs are significantly different from the sum of the contributions due to the single glucose units. Thus, a variation of the optical CD activity upon inclusion of a guest is expected, because of the conformational changes of the macrocycle and the local dipolar microenvironment effect of the guest itself [36].

2.4.2.6.5. Isothermal Titration Calorimetry (ITC)

Isothermal titration calorimetry (ITC) is a useful method for determining the thermodynamic constants, association constants and stoichiometry of interacting molecules in aqueous solutions, based on the measurement of changes in their thermodynamic parameters. ITC is presently considered one of the most interesting methods for the characterization of the interaction mechanisms of CDs with drugs. Measurement of the heat flow generated or absorbed when substances bind allows for a true measurement of the binding interactions, accurate determination of the complex stoichiometry and association constants, evaluation of enthalpy and entropy changes, thus providing in a single experiment a complete thermodynamic profile of the molecular interactions.

ITC was used to determine the thermodynamics of the complexation of hydroxypropyl-βCD with artemisinin and naproxen at different temperature and pH values; thermodynamic data revealed the involvement of hydrophobic bonding and the presence of an enthalpy–entropy compensation mechanism and provided insight on the orientation of the drug molecules inside the CD cavity. Also complexation in aqueous medium of sertaconazole with hydroxypropyl-βCD was investigated by phase-solubility analysis and ITC; the full agreement of the results obtained by the two techniques confirmed the practical interest of ITC for a fast screening of the potential of CDs as drug solubilizers [37]. The binding interaction in aqueous solution of ozonide antimalarials with βCD and sulfobutyl ether-βCD was investigated by ITC and phase-solubility analysis; all the ozonide compounds exhibited exceptionally high binding constants with both CDs, due to a very close fit within the CD cavity resulting in very favorable changes in both the enthalpy and entropy of the binding interaction [38].

2.4.3. Phthalocyanine

Phthalocyanine (Pc) (Fig. 2.9) is an intensely blue-green-coloured aromatic macrocyclic compound having p-type semiconducting properties that is widely used in dyeing [39]. Phthalocyanines form coordination complexes with most elements of the periodic table. These complexes are also intensely colored and also are used as dyes or pigments. These materials have the advantage of being sufficiently stable towards chemicals and heat. Structural, optical and electrical properties of phthalocyanine thin films have been studied extensively and they are used in nanostructured materials with a wide range of applications: photo-voltaic cell elements, nonlinear optics and solar cells. They show strong non-linear optical properties due to their spatially extended Pi-electron system. The study of phthalocyanine compounds is very important to understand the properties of their electronic behaviors under various conditions such as changes in frequency, temperature, pressure, ambient gases, etc. The properties of Pc thin films changes with various parameters such as substrate temperature, post deposition annealing, evaporation rate and thickness. Research into the behavior of phthalocyanines is normally carried out in two main ways, optically (usually absorption measurements) and electrically (using planar or sandwich devices). Planar devices consisting of thin films with inter digital electrodes are primarily used as gas sensors and to observe the materials response to various gases by DC electrical characteristics, whereas sandwich devices can be used for AC measurements of capacitance as well.

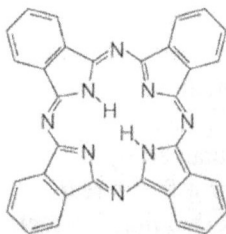

Fig. 2.9. The chemical structure of phthalocyanine, reused with permission.

2.4.3.1. Properties

Unsubstituted phthalocyanine and many of its complexes have very low solubility in organic solvents. Many phthalocyanine compounds are thermally very stable, do not melt but can be sublimed. Substituted phthalocyanine complexes often have much higher solubility. They are less thermally stable and often cannot be sublimed. Unsubstituted phthalocyanines strongly absorb light between 600 and 700 nm, thus these materials are blue or green. Substitution can shift the absorption towards longer wavelengths, changing the color from pure blue to green to colorless. Many derivatives of the parent phthalocyanine are known, where either carbon atoms of the macrocycle are exchanged for nitrogen atoms or where the hydrogen atoms of the ring are substituted by functional groups like halogens, hydroxy, amino, alkyl, aryl, thiol, alkoxy, nitro, etc.

2.4.3.2. Phthalocyanines Derivatives

The central cavity of phthalocyanines is known to be capable of accommodating 63 different elemental ions, including hydrogen (metal-free phthalocyanine, H_2-PC). A phthalocyanine containing one or two metal ions is called a metal phthalocyanine (M-PC). In the last decade, as a result of their high electron transfer abilities, M-PCs have been utilized in many fields such as molecular electronics, optical electronics, photonics, etc. The functions of M-PCs are almost universally based on electron transfer reactions because of the 18 π electron conjugated ring system found in their molecular structure.

2.4.3.3. Applications

Metallophthalocyanines (MPcs) and its derivatives exhibit many interesting properties which find application in the fields such as optical storage, electrochromic displays, photovoltaic cells and gas sensing devices due to their high chemical and thermal stability, designed flexibility, efficient electron transfer abilities, varied coordination properties, diverse substitutional alternatives and interesting electrochemical properties. Their physicochemical properties can be fine-tuned by changing the metal and/or nature of substituents. MPcs have proven to be functional species on modified electrodes especially for electrochromic devices. It is well documented that phthalocyanines are unique macromolecular complexes that catalyze many target species such as CO_2, CO, and H^+ because of their rich redox behavior. Although phthalocyanines carrying electron donating substituents have frequently been described, those with electron withdrawing groups have not been extensively studied, especially those containing fluorine atoms. Recently, several workers reported the synthesis and properties of the phthalocyanines bearing some fluoro containing substituents. Fluorine compounds are known to enhance the solubility of phthalocyanine in polar, aprotic solvents. Introducing electron-withdrawing fluorine substituents to the phthalocyanines alters the electrochemical properties of the complexes [40].

For some applications, the lower solubility of unsubstituted MPcs can present problems, but low solubility in common organic solvents can be overcome by the introduction of appropriate substituents onto the ring system. Tetrapyridoporphyrazine MPc analogues in which all four benzenoid rings are replaced by pyridinoid rings were first synthesized by Linstead and his co-workers in 1937 [41]. They obtained an insoluble product from the self-condensation of 3, 4-dicyanopyridine which was presumably a mixture of 'positional isomers' or regioisomers. Subsequently, Yokote and Shibamiya reported the synthesis and dying properties of some unsubstituted benzopyridopyridoporphyrazines [42] and the ring system attracted the attention of other groups, resulting in a substantial increase in the number of known derivatives. Yokote and Shibamiya also reported the synthesis of unsubstituted benzopyridoporphyrazines containing a mixture of benzenoid and pyridinoid rings by cross cyclotetramerization of phthalic anhydride and pyridine carboxylic anhydride [43].

Some research has been carried out in order to immobilize some compounds in microporous solid supports like silica, zeolite, alumina, carbon and polymeric matrices [44]. However, this immobilization usually results in a decrease of the catalytic activity, compared to the complexes in solution, because of the additional resistance to mass transfer when these compounds are immobilized [44]. The immobilization of these species in sol-gel matrices can produce material with application in catalysis, non-linear optics and sensors. However, the low solubility of metallo-phthalocyanines in water, can make their immobilization in solid matrices more difficult. The intermolecular interactions between the phthalocyanine rings lead to molecular aggregates during gel phase formation [45], affecting the electronic structure and consequently the electrochemical properties. All the characteristics presented can make the present electrode potentially useful as a sensor for determining oxygen in water.

2.4.3.3.1. Redox Reactions

Metal phthalocyanines have long been examined as catalysts for redox reactions. Areas of interest are the oxygen reduction reaction and the sweetening of gas streams by removal of hydrogen sulfide. Phthalocyanine compounds have been investigated as donor materials in molecular electronics, e.g. organic field-effect transistors. Copper Phthalocyanine (CuPc) may possibly be used as storage in quantum computing, due to the length of time its electrons can remain in superposition.

2.4.3.3.2. Biosensors

Metallophthalocyanines are a possible choice for preparing voltammetric modified sensors due to their catalytic activity for a wide range of redox processes. The sensitivity and the selectivity of the (bio)sensors can be greatly improved as a result of the electrocatalysis by metallophthalocyanines. For example, Oni and Nyokong [46] studied the electrocatalytic activity of iron(II) phthalocyanine complexes to DA at its modified carbon paste electrode, and also pointed out that the bulk cobalt and nickel phthalocyanine complexes did not have electrocatalytic activity to DA. The electrocatalytic behavior of these complexes towards oxidation of species is believed to be mediated by the Fe III /Fe II couple. A cation surfactant cetyltrimethylammonium bromide (CTAB) and iron(II) octanitro phthalocyanine modified carbon paste electrode was fabricated and applied to simultaneous determination of ascorbic acid, dopamine and uric acid [47]. Recently, immobilization of phthalocyanines and metallophthalocyanines at the surface of carbon nanotubes (CNTs) has been tested. CNTs promote electron transfer reactions when used as electrode materials due to their subtle electronic properties. In principle, these electrodes should show higher reversibility, low capacitive current when mixed with a proper binder, an easy renewal of the surface and finally the possibility of chemical derivatization. CNTs working electrodes have been proved beneficial helping the electron exchange reaction in redox processes of different species like DA, epinephrine, and 5-HT [48, 49]. The special properties of CNTs, together with the electrocatalytic activity of the immobilized metallophthalocyanines molecules, offer significant advantages for chemically modified electrodes development (CMEs). Since the major effect of such CMEs electrocatalysis consists of the lowering of the potential required for the electrolysis of the catalyzed redox system, these electrodes should be able to provide enhanced performance in the area of electroanalysis.

Recently, immobilization of phthalocyanines and metallophthalocyanines at the surface of carbon nanotubes has been achieved. The resulting phthalocyanine-nanotube complexes (nanocomposites) possess the catalytic properties of phthalocyanine without any destruction of the electrical properties and structures of the nanotubes and thus noncovalent functionalization of carbon nanotubes is important for developing new nanomaterials possesses a high sensitivity, fast response, ease of processability, as well as a scope of operation at room temperature; so they have therefore been studied extensively as thin films for chemical detection.

Galal et al presented a novel electrochemical sensor for morphine [50] and dobutamine [51] based on the modification of carbon paste electrode with cobalt-phthalocyanine then gold nanoparticles (AuCoPcMCPE). Gold nano-clusters and cobalt-phthalocyanine act as conductors and charge mediator centers resulting in improved electrocatalytic activity toward the studied analytes [50, 51]. In addition, Nada et al studied the effect of changing the type of metal-phthalocyanine (AuMPcMCPE) toward the electrochemical oxidation of morphine (Fig. 2.10A). AuCoPcMCPE exhibits higher current response (100 μA) and lower oxidation potential (408 mV) compared to (59.5 μA, 406 mV) at AuCuPcMCPE, (29.1 μA, 430 mV) at AuNiPcMCPE and 24.1 μA, 500 mV) at AuFePcMCPE, respectively. AuCoPcMCPE shows the ideal mediation characteristic due to the difference in the atomic size of phthalocyanine metals which resulted in variable electron transfer kinetics and difference in the reactivity between them. AuCoPcMCPE exhibits the highest current response even we change the pH of the studied drug (Fig. 2.10B) [50].

2.4.3.3.3. Gas Sensors

MPc based resistive sensors have exhibited high sensitivity towards weak concentrations of oxidizing and reducing gases such as nitrogen dioxide (NO_2), ozone (O_3), chlorine, and bromine, and potentialities to detect even volatile organic compounds. The sensing devices have been coated by means of sublimation, dip coating, spray coating, Langmuir Blodgett films. The electrical and especially gas sensing properties of MPc thin films depend on film morphology, film thickness, central metal cation, side chain etc.

84

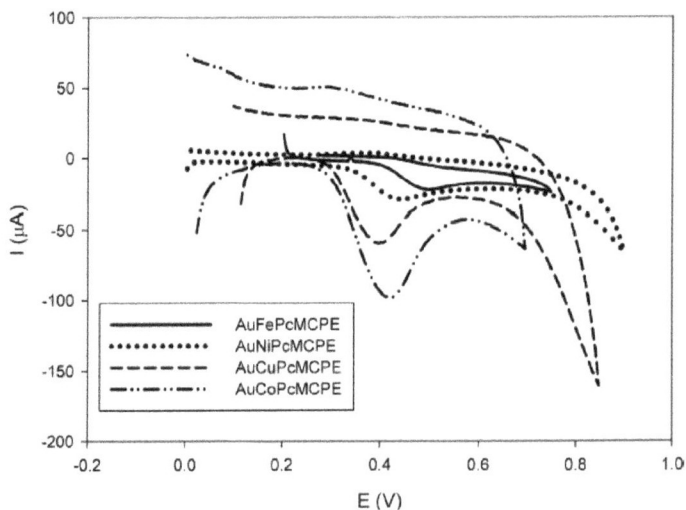

(a)

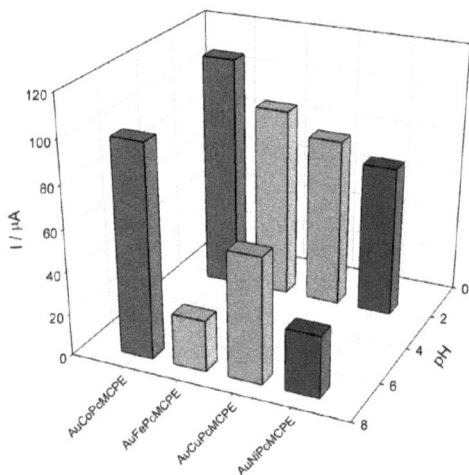

(b)

Fig. 2.10. A) Cyclic voltammograms of 1 mM Morphine in B-R buffer pH 7.4 at scan rate 100 mVs[-1] recorded at four different working electrodes: 1) AuFePcMCPE (solid line); 2) AuNiPcMCPE (dotted line); 3) AuCuPcMCPE (dashed line) and 4) Au- CoPcMCPE (dashed dotted line). B) Histogram of 1 mM Morphine at four different electrodes (AuCoPcMCPE, AuFePcMCPE, Au- CuPcMCPE and AuNiPcMCPE) in B-R buffer of pH 2 and pH 7 [50], reused with permission.

Generally, when the substituted groups are placed on the Pc ring, not only the solubility of the compound increases but supramolecular organization can be achieved, too. Mesogenity of Pc was first demonstrated in alkoxymethyl-substituted copper phthalocyanine in 1982 [52]. The aromatic ring can be considered as the core of a discotic mesogen, forming a supramolecular organization and exhibiting columnar mesophases. Then, liquid-crystalline (LC) CuPcs with different types of substituent have been synthesized and most common ones are alkyl, alkoxy or alkoxymethyl chains. The LC properties of octa-hexylthio-substituted copper (II) phthalocyanine derivatives [$(C_nS)_8PcCu$] with n = 6, 8, 10, 12, and 16 have also been investigated. It has been shown that alkylthio-substituted phthalocyanines, and the copper complexes display higher conductivities than their alkoxy-substituted phthalocyanines in their mesophase [53]. It is well known MPcs are semiconductor and have π-electron system. In general, the NO_2 sensing mechanism of MPcs has been explained in the term of a charge transfer interaction between NO_2 (electron acceptor) and MPcs (electron donor). The conductivity of MPcs is extremely increased with adsorption of NO_2, due to the hole injection after the adsorption. Due to the strong interaction between MPc and NO_2 molecule, the MPc thin film sensors are not full reversible at room temperature. In order to provide desorption of chemisorbed molecule, elevating temperature or illumination with light is needed.

Hydrogen has been considered an efficient fuel of the future because it burns cleanly and could be produced from abundant renewable energy sources. There are several methods to produce hydrogen. Molecular catalysts confined into a polymer matrix have been found to enable excellent molecular catalysis for hydrogen evolution reaction (HER). It is well documented that phthalocyanines are unique macromolecular complexes that catalyze many target species such as CO_2, CO, and H^+ because of their rich redox behavior. Carbon electrodes modified with MPc have been used for hydrogen evolution reaction (HER).

2.4.3.4. Examples of Metallophthalocyanine

2.4.3.4.1. Cobalt Phthalocyanine

Phthalocyanine complexes of transition metals, especially cobalt phthalocyanine (CoPc), are large macrocyclic molecules with applications in the fields of conductive polymers, chemical sensors, electrochromism, etc., due to their low cost facileness of large-scale preparation, and chemical and thermal stability. Numerous research groups have considered CoPc modified electrodes for determining sulfur containing compounds such as carbaryl, ethylene bis-dithio-carbamate-based pesticides, cystine, and zinc pyrithione. Cobalt phthalocyanine (CoPc) films were deposited on the surface of a screen-printed carbon electrode using a simple drop coating method. The cyclic voltammogram of the resulting CoPc modified screen-printed electrode prepared under optimum conditions shows a well-behaved redox couple due to the (Co I /Co II) system. The CoPc surface demonstrates excellent electrochemical activity towards the oxidation of sulfur [54]. However, CoPc modified electrodes were rarely utilized for cosmetic product analysis.

Cobalt phthalocyanines (CoPcs) have been known for a long time as electrocatalysts for the oxygen reduction reactions (ORR). Mho et al. reported the interactions of reduced CoPcs with oxygen and proposed that the catalytic activities of phthalocyanines are achieved through a conventional regenerative catalytic mechanism [55].

Advantage of the electrocatalytic activity of a CoPcs modified carbon paste electrode for citric acid oxidation towards developing a very straight forward amperometric flow injection analysis method well suited for the analysis of commercial citric juices with minimum sample pretreatment. Ascorbic acid interference is circumvented by sample pretreatment with ascorbate oxidase. Recently, immobilization of phthalocyanines and metallophthalocyanines at the surface of carbon nanotubes has been achieved. The resulting phthalocyanine-nanotube complexes possess the catalytic properties of phthalocyanine without any destruction of the electrical properties and structures of the nanotubes and thus noncovalent functionalization of carbon nanotubes is important for developing new nanomaterials. Although CoPMCPEs have been used for determination of organic compounds such as phenols and dithiocarbamate fungucides, sensitive electrochemical detection of hydrogen peroxide can be achieved at BDD electrodes modified with CoPc [56]. Thus the use of CoPc-BDD should be effective for fabrication of glucose oxidase-modified BDD electrodes for sensitive glucose detection. It has been observed that phthalocyanine–CNT complexes have excellent catalytic properties of phthalocyanines without losing any of the electronic properties of carbon nanotubes. Dopamine sensors made from layered double hydroxide nanosheets and cobalt phthalocyanines have been created by a team of Chinese scientists. The modified electrode exhibits a low detection limit, fast response and long - term stability for the determination of dopamine. Recent work conducted by Moraes et al has led to the development of an electrochemical device using phthalocyanine-functionalized carbon nanotubes as a sensor for dopamine contaminated with a high concentration of ascorbic acid, reaching a detection limit of 2.6×10^{-7} mol L^{-1} [57] and for selective determination of epinephrine in urine [58].

2.4.3.4.2. Iron Phthalocyanine

If iron (II) phthalocyanine is deposited at room temperature, the grains are small and round. The advance of thin film deposition of organics allows a control of the structural quantities, such as film thickness, grain size, and molecule orientation with respect to the substrate. For example, two parameters to control the structure of phthalocyanines are the substrate temperature during deposition and the type of substrate the phthalocyanine is sublimed onto. The resulting film morphology often impacts the electronic properties. It was found that chemical sensors for thin films have a superior chemical sensitivity than thicker films. The chemical response to various analytes (methanol, DMMP, H_2O_2, nitro benzene, etc.) is measured in a flow system with precise temperature and flow concentration control.

(FePc) modified multi-wall carbon nanotubes paste electrodes (MWCNTPEs) were used as voltammetric sensors to selectively detect dopamine (DA). For example, Oni and Nyokong [46] studied the electrocatalytic activity of iron(II) phthalocyanine complexes to DA at its modified carbon paste electrode, and also pointed out that the bulk cobalt and nickel phthalocyanine complexes did not have electrocatalytic activity to DA. The electrocatalytic behavior of these complexes towards oxidation of species is believed to be mediated by the FeIII/FeII couple. Also cation surfactant cetyltrimethylammonium bromide (CTAB) and iron(II) octanitro phthalocyanine modified carbon paste electrode was fabricated and applied to simultaneous determination of ascorbic acid, dopamine and uric acid [47].

Iron phthalocyanines (FePcs) have shown good electrocatalytic behaviour for the detection of analytes such as hydrazine, amitrole, and l-cysteine. The attachment of MPcs to an electrode using click chemistry is a point of interest. Although the synthesis of MPcs bearing terminal alkyne and azide groups is possible, it is often complex and commonly requires the use of protecting groups. A different approach, relying on the strong axial ligation of iron to pyridine, was recently investigated and showed potential for the detection of hydrazine as a test analyte. It is well understood that FePc forms strong axial bonds to pyridine, through the coordination to the nitrogen base by the metal center. It has also been reported that iron phthalocyanine can be attached to self-assembled monolayers of 4-mercaptopyridine on gold [59]. In order to avoid the synthesis of elaborate and expensive MPcs, the use of an absorbed layer of SWCNTs, which are grafted with 4-azidobenzenediazonium salts and then linked to commercially available 4-ethynylpyridine by click chemistry was investigated. Thereafter, unsubstituted FePc was bound to this surface through the strong axial bond with pyridine.

2.5. Conclusion

As conclusion, each modifier has their own additives as modifier agent. There are many modifier materials that can be used to mix with substituent to improve their performance, such as lowering their working potential to reduce interference and exhibit much faster electron transfer so as to give higher sensitivity. They also exhibit fast reversible electrochemical response at negative potentials, which makes them useful as redox mediators for numerous enzymatic reactions. Nevertheless, the selection of modifier is depending on the user or company.

References

[1]. S. Tajik, M. A. Taher, H. Beitollahi, Simultaneous determination of droxidopa and carbidopa using a carbon nanotubes paste electrode, *Sensors and Actuators B*, Vol. 188, 2013, pp. 923–930.

[2]. R. W. Murray, J. B. Goodenough, W. J. Albery, Modified Electrodes: Chemically Modified Electrodes for Electrocatalysis, *The Royal Society*, 1981, pp. 253-265.

[3]. http://virginia.academia.edu/BankimSanghavi

[4]. A. F. Neto, A. C. Pelegrino, V. A. Darin, Ferrocene: 50 Years of Transition Metal Organometallic Chemistry – From Organic and Inorganic to Supramolecular Chemistry, *ChemInform*, Vol. 35, 2004, pp. 43.

[5]. S. Coriani, A. Haaland, T. Helgaker, P. Jørgensen, The Equilibrium Structure of Ferrocen, *Chemical Physics and Physical Chemistry*, 2006, Vol. 7, pp. 245–249.

[6]. Tai S. Chao et al., Iron-containing motor fuel compositions and method for using same, *U. S. Patent 4,104,036,* 1978.
[7]. W. Y. Liu, Q. H. Xu, Y. X. Ma, Y. M. Liang, N. L. Dong, D. P. Guan, Solvent-free synthesis of ferrocenylethene derivatives, *Journal of Organometallic Chemistry*, Vol. 625, Issue 1, 2001, pp. 128–132.
[8]. T. Kuwana, W. G. French, Carbon paste electrodes containing some electroactive compounds, *Analytical Chemistry*, Vol. 36, Issue 1, 1964, pp. 241-242.
[9]. N. F. Atta, A. Galal, A. A. Wassel, A. H. Ibrahim, Sensitive Electrochemical Determination of Morphine Using Gold Nanoparticles–Ferrocene Modified Carbon Paste Electrode, *International Journal of Electrochemical Science,* Vol. 7, 2012, pp. 10501–10518.
[10]. N. F. Atta, A. Galal, S. M. Ali, S. H. Hassan, Electrochemistry and detection of dopamine at a poly(3, 4-ethylenedioxythiophene) electrode modified with ferrocene and cobaltocene, *Ionics*, Vol. 21, 2015, pp. 2371-2382.
[11]. N. F. Atta, A. Galal, S. M. Ali, S. H. Hassan, Electrochemical Sensor for Morphine Based on Gold Nanoparticles/ Ferrocene Carboxylic Acid/Poly (3,4-Ethylene-Dioxythiophene) Composite, *International Journal of Electrochemical Science*, Vol. 10, 2015, pp. 2265–2280.
[12]. Y. Lu, X. Li, L. Zhang, P. Yu, L. Su, L. Mao, Aptamer-based electrochemical sensors with aptamer-complementary DNA oligonucleotides as probe, *Analytical Chemistry*, Vol. 80, Issue 6, 2008, pp. 1883–1890.
[13]. M. Pawlak, E. Grygolowicz-Pawlak, E. Bakker, Ferrocene bound poly(vinyl chloride) as ion to electron transducer in electrochemical ion sensors, *Analytical Chemistry*, Vol. 82, Issue 16, 2010, pp. 6887–6894.
[14]. B. Ermolinsky, M. Peredelchuk, D. Provenzano, Alpha cyclodextrin decreases cholera toxin binding to GM1 gangliosides, *Journal of Medical Microbiology*, Vol. 62, 2013, pp. 1011-1014.
[15]. K. Uekama, F. Hirayama, T. Irie, Cyclodextrin drug carrier systems, *Chemical Reviews*, Vol. 98, Issue 5, 1998, pp. 2045–2078.
[16]. J. Szejtli, Introduction and general overview of cyclodextrin chemistry, *Chemical Reviews*, Vol. 98, Issue 5, 1998, pp. 1743–1753.
[17]. S. K. Rodal, G. O. Grethe, F. Vilhardt, B. van Deurs, K. Sandvig, Extraction of Cholesterol with Methyl-β-Cyclodextrin Perturbs Formation of Clathrin-coated Endocytic Vesicles, *Molecular Biology of the Cell,* Vol. 10, Issue 4, 1999, pp. 961–74.
[18]. L. Szente, J. Szejtli, Highly soluble cyclodextrin derivatives: chemistry, properties, and trends in development, *Advanced Drug Delivery Reviews,* Vol. 36, Issue 1, 1999, pp. 17–28.
[19]. T. R. Thatiparti, A. J. Shoffstall, H. A. Von Recum, Cyclodextrin-based device coatings for affinity-based release of antibiotics, *Biomaterials,* Vol. 31, Issue 8, 2010, pp. 2335–2347.

[20]. R. A. Rajewski, V. J. Stella, Pharmaceutical applications of cyclodextrins, In vivo drug delivery, *Journal of Pharmaceutical Sciences,* Vol. 85, Issue 11, 1996, pp. 1142–1168.

[21]. N. F. Atta, A. Galal, S. M. Ali, D. M. El-Said, Improved host–guest electrochemical sensing of dopamine in the presence of ascorbic and uric acids in a b-cyclodextrin/Nafion®/polymer nanocomposite, *Analytical Methods*, Vol. 6, 2014, pp. 5962-5971.

[22]. N. F. Atta, A. Galal, D. M. El-Said, A Novel Electrochemical Sensor for Paracetamol Based on β-Cyclodextrin/Nafion/Polymer Nanocomposite, *International Journal of Electrochemical Science,* Vol. 10, 2015, pp. 1404–1419.

[23]. N. F. Atta, E. H. El-Ads, Y. M. Ahmed, A. Galal, Determination of some neurotransmitters at cyclodextrin/ionic liquid crystal/graphene composite electrode, *Electrochimica Acta*, Vol. 199, 2016, pp. 319-331.

[24]. N. F. Atta, S. S. Elkholy, Y. M. Ahmed, A. Galal, Host Guest Inclusion Complex Modified Electrode for the Sensitive Determination of a Muscle Relaxant Drug, *Journal of the Electrochemical Society*, Vol. 163, 7, 2016, pp. B403-B409.

[25]. L. Leclercq, H. Bricout, S. Tilloy, E. Monflier, Biphasic aqueous organometallic catalysis promoted by cyclodextrins: Can surface tension measurements explain the efficiency of chemically modified cyclodextrins, *Journal of Colloid and Interface Science,* Vol. 307, Issue 2, 2007, pp. 481–487.

[26]. A. Motoyama, A. Suzuki, O. Shirota, R. Namba, Ryujiro, Direct determination of pindolol enantiomers in human serum by column-switching LC-MS/MS using a phenylcarbamate-β-cyclodextrin chiral column, *Journal of Pharmaceutical and Biomedical Analysis*, Vol. 28, Issue 1, 2002, pp. 97–106.

[27]. J. M. Madrid, M. Villafruela, R. Serrano, F. Mendicuti, Experimental thermodynamics and molecular mechanics calculations of inclusion complexes of 9-methyl antracenoate and 1-methyl pyrenoate with beta-cyclodextrin, *Journal of Physical Chemistry B,* Vol. 103, Issue 23, 1999, pp. 4847–4853.

[28]. L. Fielding, Determination of association constants (Ka) from solution NMR data, *Tetrahedron*, Vol. 56, Issue 34, 2000, pp. 6151–6170.

[29]. C. Garnero, M. Longhi, Study of ascorbic acid interaction with hydroxypropyl--cyclodextrin and triethanolamine, separately and in combination, *Journal of Pharmaceutical and Biomedical Analysis,* Vol. 45, Issue 4, 2007, pp. 536–545.

[30]. M. S. Ibrahim, I. S. Shehatta, A. A. Al-Nayeli, Voltammetric studies of the interaction of lumazine with cyclodextrins and DNA, *Journal of Pharmaceutical and Biomedical Analysis,* Vol. 28, Issue 2, 2002, pp. 217–225.

[31]. E. Junquera, M. Martin-Pastor, E. Aicart, Molecular encapsulation of flurbiprofen and/or ibuprofen by hydroxypropyl-cyclodextrin aqueous

solution, Potentiometric and molecular modeling studies, *Journal of Organic Chemistry*, Vol. 63, Issue 13, 1998, pp. 4349–4358.

[32]. E. Junquera, J. C. Romero, E. Aicart, Behavior of tricyclic antidepressants in aqueous solution: self-aggregation and association with cyclodextrin, *Langmuir*, Vol. 17, Issue 6, 2001, pp. 1826–1832.

[33]. D. W. Armstrong, F. Nome, L. A. Spino, T. D. Golden, Efficient Detection, Evaluation of cyclodextrin multiple complex formation, *Journal of American Chemical Society*, Vol. 108, 1996, pp. 1418–1421.

[34]. J. P. Hummel, W. J. Dreyer, Measurement of protein-binding phenomena by gel filtration, *Biochimica et Biophysica Acta,* Vol. 63, Issue 3, 1962, pp. 530–532.

[35]. V. R. Sinha, R. Anitha, S. Ghosh, A. Nanda, R. J. Kumria, Complexation of celecoxib with beta-cyclodextrin: characterization of the interaction in solution and in solid state, *Journal of Pharmaceutical Science,* Vol. 94, Issue 3, 2005, pp. 676–687.

[36]. P. Lo Meo, F. D'Anna, S. Riela, M. Gruttadauria, R. Noto, Polarimetry as a useful tool for the determination of binding constants between cyclodextrins and organic guest molecules, *Tetrahedron Letter*, Vol. 47, Issue 51, 2006, pp. 9099–9102.

[37]. A. I. R. Perez, C. R. Tenreiro, C. A. Lorenzo, P. Taboada, A. Concheiro, J. J. T. Labandeira, Sertaconazole/hydroxypropyl--cyclodextrin complexation: isothermal titration calorimetry and solubility approaches, *Journal of Pharmaceutical Science,* Vol. 95, Issue 8, 2006, pp. 1751–1762.

[38]. C. S. Perry, S. A. Charman, R. J. Prankerd, F. C. Chiu, M. J. Scanlon, D. Chalmers, W. N. Charman, The binding interaction of synthetic ozonide antimalarials with natural and modified beta-cyclodextrins, *Journal of Pharmaceutical Science,* Vol. 95, Issue 1, 2006, pp. 146–158.

[39]. T. P. Hulser, H. Wiggers, F. E. Kruis, A. Lorke, Nano structured gas sensors and electrical characterization of deposited SnO2 nanoparticles in ambient gas atmosphere, *Sensors and Actuators B*, Vol. 109, Issue 1, 2005, pp. 13–18.

[40]. I. Acar, Z. Bıyıklıoglu, A. Koca, H. Kantekin, *Polyhedron,* Vol. 29, Issue 5, 2010, pp. 1475-1484.

[41]. R. P. Linstead, E. G. Noble, J. M. Wright, Phthalocyanines IX, Derivatives of thiophene thionaphthene, pyridine and pyrazine, and a note on the nomenclature, *Journal of the Chemical Society,* Vol. 11, Issue 1, 1937, pp. 911-921.

[42]. M. Yokote, F. Shibamiya, S. Shoji, Copper tetra-3,4-pyridinoporphyrazine from pyridine-3, 4-dicarboxylic acid, *Journal of the Chemical Society Japan, Industrial Chemistry Section,* Vol. 67, Issue 12, 1964, pp. 166-168.

[43]. M. Yokote, F. Shibamiya, Cu Bz-azaphthalocyanines. *Journal of the Chemical Society Japan, Industrial Chemistry Section*, Vol. 62, 1959, pp. 224-227.

[44]. G. Langhendries, G. V. Baron, P. E. Neys, P. A. Jacobs, *Chemical Engineering Science*, Vol. 54, Issue 15-16, 1999, pp. 3563- 3568.

[45]. H. Xia, M. Nogami, Copper phthalocyanine bonding with gel and their optical properties, *Optical Materials*, Vol. 15, Issue 2, 2000, pp. 93-98.

[46]. J. Oni, T. Nyokong, Simultaneous voltammetric determination of dopamine and serotonin on carbon paste electrodes modified with iron(II) phthalocyanine complexes, *Analytica Chimica Acta*, Vol. 434, Issue 1, 2001, pp. 9–21.

[47]. R. R. Naik, E. Niranjana, B. E. K. Swamy, B. S. Sherigara, H. Jayadevappa, Surfactant induced iron (II) phthalocyanine modified carbon paste electrode for simultaneous detection of ascorbic acid, dopamine and uric acid, *International journal of Electrochemical Science*, Vol. 3, 2008, pp. 1574–1583.

[48]. R. Antiochia, I. Lavagnini, F. Magno, F. Valentini, G. Palleschi, Single-wall carbon nanotube paste electrodes: a comparison with carbon paste, platinum and glassy carbon electrodes via cyclic voltammetric data, *Electroanalysis,* Vol. 16, Issue 17, 2004, pp. 1451–1458.

[49]. F. Valentini, S. Orlanducci, E. Tamburri, M. L. Terranova, A. Curulli, G. Palleschi, Single-walled carbon nanotubes on tungsten wires: a new class of microelectrochemical sensors, *Electroanalysis,* Vol. 17, Issue 1, 2005, pp. 28–37.

[50]. N. F. Atta, A. Galal, F. M. Abdel-Gawad, E. F. Mohamed, Electrochemical Morphine Sensor Based on Gold Nanoparticles Metalphthalocyanine Modified Carbon Paste Electrode, *Electroanalysis*, Vol. 26, 2014, pp. 1–15.

[51]. N. F. Atta, A. Galal, F. M. Abdel-Gawad, E. F. Mohamed, Electrochemistry and Detection of Dobutamine at Gold Nanoparticles Cobalt-Phthalocyanine Modified Carbon Paste Electrode *Journal of the Electrochemical Society*, Vol. 162, 12, 2015, pp. B304-B311.

[52]. C. Piechocki, J. Simon, A. Skoulios, D. Guillon, P. Weber, Discotic mesophases obtained from substituted metallophthalocyanines. Toward liquid crystalline one-dimensional conductors, *Journal of the American Chemical Society*, Vol. 104, Issue 19, 1982, pp. 5245–5247.

[53]. K. Z. Ban, K. Nishizawa, K. Ohta, H. Shirai, Discotic liquid crystals of transition metal complexes 27: supramolecular structure of liquid crystalline octakisalkylthiophthalocyanines and their copper complexes, *Journal of Materials Chemistry,* Vol. 10, Issue 5, 2000, pp. 1083–1090.

[54]. C. Pei-Yen, L. Chin-Hsiang, C. Mei-Chin, T. Feng-Jie, C. Nai-Fang, S. Ying, Screen-Printed Carbon Electrodes Modified with Cobalt Phthalocyanine for Selective Sulfur Detection in Cosmetic Products, *International Journal of Molecular Science*, Vol. 12, Issue 6, 2011, pp. 3810-3820.

[55]. S. Mho, B. Ortiz, N. Doddapaneni, S. M. Park, Electrochemical and Spectroelectrochemical Studies on Metallophthalocyanine-Oxygen Interactions in Nonaqueous Solutions, *Journal of the Electrochemical Society*, Vol. 142, Issue 4, 1995, pp. 1047-1053.

[56]. T. Kondo, A. Tamura, and T. Kawai, Cobalt phthalocyaninemodified boron-doped diamond electrode for highly sensitive detection of hydrogen

peroxide, *Journal of the Electrochemical Society*, Vol. 156, Issue 11, 2009, pp. F145–F150.

[57]. F. C. Moraes, M. F. Cabral, S. A. S. Machado, L. H. Mascaro, Electrocatalytic behavior of glassy carbon electrodes modified with multiwalled carbon nanotubes and cobalt phthalocyanine for selective analysis of dopamine in presence of ascorbic acid, *Electroanalysis*, Vol. 20, Issue 8, 2008, pp. 851–857.

[58]. C. F. Moraes, D. L. C. Golinelli, L. H. Mascaro, S. A. S. Machado, Determination of epinephrine in urine using multi-walled carbon nanotube modified with cobalt phthalocyanine in a paraffin composite electrode, *Sensors and Actuators B*, Vol. 148, Issue 2, 2010, pp. 492–497.

[59]. K. I. Ozoemena, T. Nyokong, Electrocatalytic oxidation and detection of hydrazine at gold electrode modified with iron phthalocyanine complex linked to mercaptopyridine self-assembled monolayer, *Talanta*, Vol. 67, Issue 1, 2005, pp. 162-168.

3.

Self-Assembly Monolayers: New Strategy of Surface Modification for Sensor Applications

Nada F. Atta, Ekram H. El-Ads and Ahmed Galal

3.1. General Introduction

Currently, covalently modified surfaces have received much attention as they proved very effective in various applications such as material science, biophysics, corrosion protection, optical devices, electrocatalysis, integrated circuits, electroanalysis, information storage, sensing and biosensing. Self-assembling of monolayers over the substrate surface is an important approach that has been considered recently presenting high control of modified surfaces. Surface can be decorated with mono or multi-functionalities resulting in the formation of unique or mixed monolayers of molecular level. This approach is very pretty for sensing and biosensing applications due to the requirement of particular interactions with the studied species in solution [1, 2]. SAMs (Self-Assembled Monolayers) present several effective advantages such as high stability and endurance, easy preparation and molecular control therefore they appear as a "shining star" in the field of surface modification. SAMs are formed through the spontaneous adsorption (physical or chemical) of organic compounds over the surface of the substrate resulting in the formation of highly ordered single molecular monolayer (Fig. 3.1) [3]. There are attractive forces between the particular functional groups in the adsorbed organic compound and the substrate surface [1, 4].

Self-assembling of organothiols over the gold surface is an effective and pretty approach in the chemically modified electrodes development [5]. The best studied and the most characterized SAMs available for sensor applications are those resulted from the chemisorption of alkanethiols from solution over the gold surface. Thiols are the anchoring sites of the compounds onto the gold surface due to the high binding affinity between sulfur and gold which results in unique SAMs. They tender many advantages like densely packed assemblies, highly organized system, easy preparation and the potential of proffer wide varieties of effective functional groups at the SAMs surface. Also, they can present a highly ordered monolayer of molecular level and can engineer the energy of the desired electrode surfaces. In addition, they can monitor the interfaces properties (chemical and physical) for wide range of applications. Therefore, they are suitable systems for sensor applications due to their unique features [2-5]. SAMs of functionalized organothiols are suitable systems for studying the molecular interactions, molecular recognition and other phenomena in biological systems that depend on pH. Also, they are used for proteins immobilization and have potential roles in designing electrochemical sensors and biosensors [5].

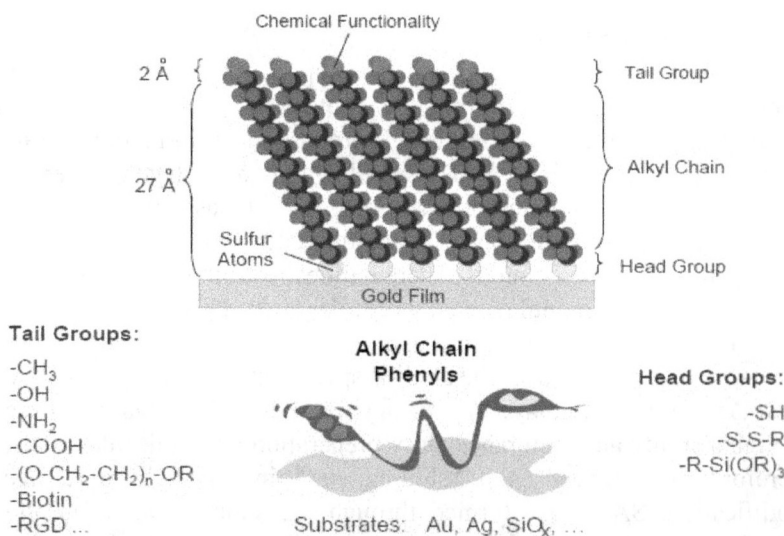

Fig. 3.1. Schematic graph of SAM (S-containing compounds) formation on the substrate, reused with permission [3].

3.1.1. SAM Structure

The self-assembly structure is greatly affected by the SAM formation method, electrode material choice and redox center attachment. The formation of well-ordered SAMs through the spontaneous adsorption of thiols and disulfides on Au was firstly reported in 1980. A SAM is depicted as well-uniformly ordered molecules on a flat surface. But there is a wide variety of defects are possible in the monolayer like domains, islands, collapsed sites "disorder" and pinholes "molecular vacancies" [6].

It was assumed that the metallic surface for the SAM formation is flat but there is a wide variety of features like steps, crystalline boundaries and terraces were observed by different scanning techniques. Therefore, it is very important to select the electrode material which affected the SAM ordering and integrity. Mercury, as a liquid, formed featureless and defect-free surfaces while gold, silver and other metals, as polycrystalline surfaces, contained atomic scale defects resulted in the complication of SAM formation. Also, it was assumed that the organic molecules of the SAM are uniformly aligned on the surface.

The monolayer packing was significantly varied by the tilt angle of the alkane thiols. Alkane thiols lay flat across the surface at the early stages of SAM formation and by time most of the SAM molecules order with the molecular axis normal to the surface. Some of the chains are still flat on the surface. The molecular vacancies in the SAMs resulted in the formation of pinholes [6]. Defects in the form of pinholes were considered electrochemically as ultramicroelectrodes array. A SAM with pinholes defects "disordered SAM" allows the approach of the electrolyte to the electrode surface resulting in the increment of the capacitive current. The possible SAM defects may be in the form of pinholes, disorder, domains and islands. The effect of varying the chain length of hydroxyalkane thiol SAMs (from 2 to 16 methylene units) on the capacitance value was examined (using cyclic voltammetry). Upon increasing the methylene unit number, the capacitance decreased from 12.6 $\mu F/cm^2$ (in case of 2 methylene units) to 1.36 $\mu F/cm^2$ (in case of 16 methylene units). This observation indicated that the pinholes defects were disappeared in the SAMs [6].

3.1.2. Different Types of SAMs

3.1.2.1. SAM of Organothiols

SAMs based on alkane thiols on gold (Au) are the most common type of SAMs due to their ease of preparation and stability. Long-chain alkane thiols are more preferred in forming highly ordered and well-packed

SAMs compared to short alkane thiols. The van der Waals interactions through the SAMs are stronger in case of long-chain alkane thiols and weaker in case of short chain alkane thiols [6]. Alkane thiol molecules can form stable and chemically bound SAMs on the surface of Au, Ag or Hg [7]. One of the most common alkane thiols that can form highly stable SAMs on gold and Pt for sensing applications is cysteine ($HSCH_2CHNH_2COOH$). Cysteine contains amino and carboxylic acid functional groups useful for biomolecules conjugation like enzyme, DNA and antibodies [4]. Besides, SAMs based on aromatic and heteroaromatic thiols/disulfides are investigated recently because of the stronger intermolecular interactions than the case of alkanethiols. This results in different molecular packing assemblies. In addition, the SAMs based on aromatic thiols exhibit higher conductance compared to alkane thiols due to the delocalization of electros in the benzene ring [5, 8]. Examples of organothiols molecules (alkanethiol, arenethiol, alkanedithiol, dialkyldisulfide and dialkylsulfide) that can be adsorbed on different substrates are shown in Fig. 3.2 [9]. Anisotropic aromatic and heteroaromatic thiols have the affinity to form SAMs on metal surfaces like mercaptobenzimidazole, mercaptobenzothiazole, mercaptoimidazole, mercaptopyrimidine and thiophenol [8]. Thiols can be used to modify the surface for sensors/biosensors applications through different approaches. Fig. 3.3 shows five approaches for SAM modification involve 1) adsorption of functionalized thiols, 2) functionalization then adsorption, 3) adsorption then functionalization, 4) adsorption of mixed SAMs composed of two or more different thiols and 5) adsorption then incorporation [10].

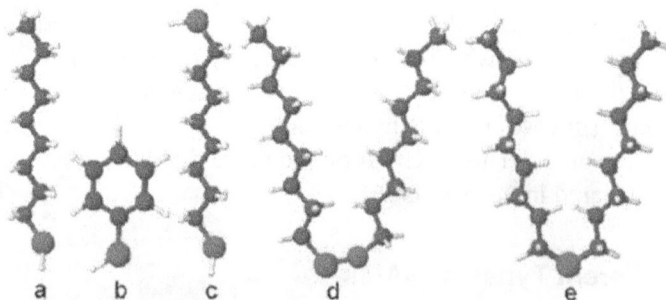

a   b   c   d   e

Fig. 3.2. Some examples of sulfur compounds that form self-assembled monolayers on metals and semiconductors: (a) alkanethiol (nonanethiol); (b) arenethiol (benzenethiol); (c) alkanedithiol (octanedithiol); (d) dialkyldisulfide (dinonyl disulfide); (e) dialkylsulfide (dinonyl sulfide). Red: S atom, blue: C atom, white: H atom, reused with permission [9].

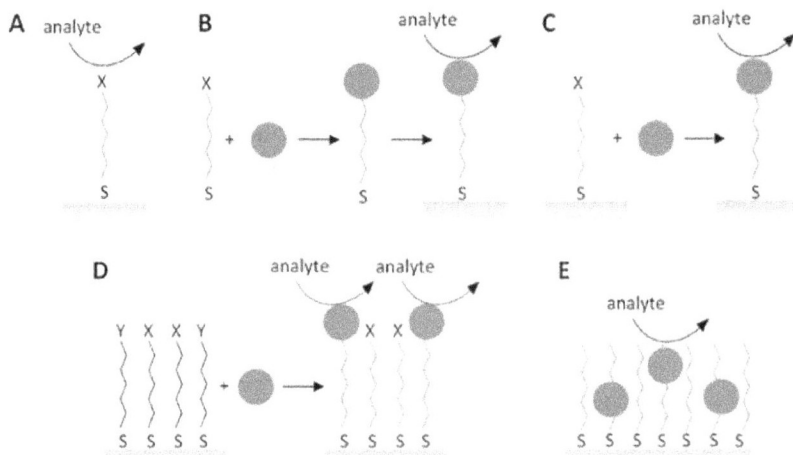

Fig. 3.3. Schematics of the different approaches for assembling thiol-based SAMs. a Functionalized thiols, b functionalization followed by attachment, c attachment followed by functionalization, d attachment of a mixed layer, and e attachment followed by incorporation, reused with permission [10].

3.1.2.2. SAMs of Phosphates

SAMs of alkane-phosphates and phosphonates are well-formed on oxide surfaces like TiO_2, Ta_2O_5 and Nb_2O_5. The molecular structure of phosphates SAMs are similar to the SAMs of thiols on Au with an intermolecular spacing of 5 Å and a tilt angle of the molecular axis of 30-35° relative to the surface normal. A direct coordination between phosphate and metal cation is the basis of binding of phosphate head group to the metal oxide surface. Mono- and bi-dentate binding of phosphate group resulted in the formation of a closely and densely packed SAMs [11, 12].

3.1.2.3. SAMs of Phthalocyanines

Phthalocyanines are macromolecules that can be tailored to form SAMs or multilayer ultrathin films of tens or hundreds of nanometers thickness. They can be covalently bonded to the surface forming SAM that can be used as the sensitive part in quartz crystal microbalance. The phthalocyanine SAM can be linked to Au substrate through thiol binder or sulfur bridge atom [13].

3.1.2.4. SAMs of Silanes

Silanes and their derivatives (like n-octadecyltrichlorosilane) can be adsorbed on the surface of Si wafers forming covalently densely packed SAMs. The SAMs of silanes can be formed through the arrival of silanes molecules to the surface followed by hydrolysis then covalent grafting. The Si-Cl bond is hydrolyzed forming silanol groups (Si-OH) consuming three moles of water for each mole of n-alkyltrichlorosilane. Then the silanol groups formed cross-linked structure together and/or with the surface. Another approach is achieved through:

i) Physical two-dimensional ordering: the hydrolyzed molecules are adsorbed on the surface and organize themselves.

ii) Chemical grafting: the organized film is covalently linked to the surface [12].

3.1.2.5. SAMs of Carbenes and Diazonium Salts

Carbenes and N-heterocyclic derivatives (NHC) can form SAMs on gold substrate demonstrating greater resistance toward chemical reagents and heat than the SAMs based on thiol. SAMs of carbenes are stable under different conditions of high temperature, organic solvents, boiling water, pH extremes, etc. This type of SAM shows greater stability due to the higher strength of gold–carbon bond compared to that of S-Au bond and different mode of bonding for carbon based SAMs. The Au–NHC bond is stronger than that of Au-phosphine bond and Au-S bond resulting in ultra-stable SAMs based on NHC [14]. In addition, SAMs based on diazo compounds show increased stability [14]. Gold electrode can be modified with 4-carboxyphenyl diazonium salts resulting in the formation of stable SAMs for sensing applications. This type of SAM can withstand repeated cycling and the applied potential window. Also, it exhibits longer term stability and higher repeated cyclic stability than SAM based on alkanethiols in terms of applied extremes (electrode potential, time and sonication) [2].

3.1.3. Different Types of Substrates for SAMs

Gold and silver are the most widely used substrates for SAMs based on thiols. Other metallic substrates have been investigated for SAMs formation like nickel, platinum, palladium, copper, mercury [6], zinc, aluminum, iron, silver and titanium [1, 15]. Thiols exhibited a great affinity for the as mentioned metals resulting in the formation of highly densely packed monolayers. Also, some semiconductors [6] and nonmetallic substrates [15] have been recently investigated for SAMs formation like GaAs, InP, ITO and ZnSe.

3.1.3.1. Gold, Platinum and Glassy Carbon

Organosulfur compounds are chemisorbed spontaneously on gold surface, lose the hydrogen from the thiol group as molecular hydrogen H_2 resulting in a strong, covalent and thermodynamically favored S-Au bond formation. Gold substrate is preferred for adsorbing thiols due to i) the inertness of gold metal so that it doesn't form any stable oxides on its surface and ii) strong specific interaction of Au with sulfur allowing the formation of stable reproducible SAMs in a very short time. Selectivity, sensitivity, small overpotential and fast response time in electrocatalytic reactions are the features offered by the stable, densely-packed and well-organized SAMs of thiols on Au. SAMs can be formed on Au, Pt and GC. Galal et al investigated the impact of SAMs formation of cysteine on Au, Pt and GC. They investigated the response of SAMs modified Au, Pt or GC toward dopamine (DA) oxidation (Fig. 3.4A). They concluded that the impact of cysteine as a mediator and a bridge component for electron transfer is more obvious in case of Au substrate. This is attributed to the strong affinity of Au to S-containing compounds. In addition, there is a partial interaction between Pt and cysteine as observed from the shift of the oxidation peak potential of dopamine from 680 mV at bare Pt to 653 mV in case of Cys-Pt accompanied by the appearance of the reduction peak at 356 mV. On the other hand, the effect of cysteine on GC is less pronounced. The oxidation and reduction peak of dopamine shifted from 738 and 236 mV at bare GC to 700 and 269 mV at Cys-GC, respectively. Also, the separation of binary mixture of ascorbic acid (AA) and DA at Cys-Au was achieved well with two well-separated oxidation peaks at 402 and 606 mV, respectively (Fig. 3.4B). A combined overlapped peak was obtained in case of Cys-GC at 731 mV. At Cys-Pt, two oxidation peaks are obtained which are more resolved than in case of GC confirming the partial interaction between cysteine and Pt [16, 17].

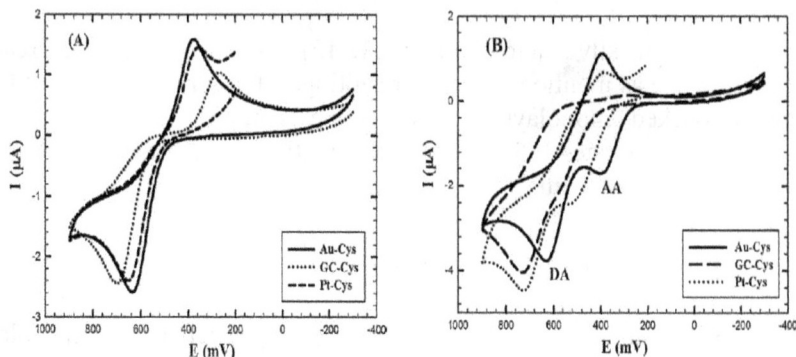

Fig. 3.4. (A) CVs of 1 mmol L^{-1} DA at Au-Cys, Pt-Cys, and GC-Cys electrodes, (B) CVs of binary mixture of 1 mmol L^{-1} AA and 1 mmol L^{-1} DA at Au-Cys, GC-Cys, and Pt-Cys electrodes, scan rate 50 mV s^{-1}, reused with permission [16].

3.1.3.2. Mercury

Mercury exhibits the characteristic of the formation of featureless surface which is free from defects as it exhibits the liquid state at the room temperature. It shows higher affinity towards thiols compared to other metals resulting in the formation of tightly packed monolayers. These tightly packed monolayers lead to the blocking of both the hydrophilic and hydrophobic redox probes like $Fe(CN)_6^{3-}$ or $Ru(NH_3)_6^{3+}$ [6].

3.1.3.3. Silver and Copper

Well-ordered SAMs are formed in case of Cu and Ag under an inert atmosphere. The tilt angle of alkane thiols on Cu and Ag substrates is smaller than that in case of Au. Also, no odd-even effects of SAMs on electron transfer rate have been observed. Upon exposure to air, an oxide film is formed on Cu and Ag. Then SAMs of alkanethiols are formed on the oxidized surfaces which are different in structure and properties from that formed on less oxidized surfaces. Then a redox reaction occurs between the metal oxide film and the alkane thiols SAMs. The molecules of alkanethiols SAMs are oxidized into sulfonates and silver and copper oxides are reduced into copper and silver, respectively. Following the reduced metal oxides react again with alkanethiols molecules to form SAMs [6]. In addition, B. Arezki investigated the emission processes of molecules-metal cluster ions from SAMs of octanethiols on Au and Ag. The resulted spectra from SAM/Ag are more intense than that in case of SAM/Au. This is due to the difference in the electronegativity between metal (Au and Ag) and S resulting in a more ionic Ag-S bond [18].

3.1.3.4. Nickel

Homogenous SAMs of thiols are formed on Ni as confirmed by Auger electron spectroscopy, electrochemical surface coverage and XPS. Ni tends to form an oxide film at ambient pressure and temperature. This oxide film must be removed via electroreduction step followed by the SAM formation by immersing in the alkane thiols solution. Some oxides may be formed due to the exposure to atmosphere between the two steps. Also, the electroreduction and the SAM formation on Ni surfaces can be achieved simultaneously using alkane thiols saturated basic solutions [6].

3.1.3.5. Palladium

Palladium and its alloys show great impact in electronic devices; as components in capacitors, resistors and electrodes in integrated circuits. Pd also serves as a main component in hydrogen gas sensor due to its easily dissociation and high absorption up to 900 times its volume in hydrogen. It is used as a superior catalyst for metals electroless deposition and for several reactions such as hydrogenation and hydrodesulfurization [19]. The alkane chains of SAMs on crystalline Pd are conformationally disordered in case of short chain lengths and they are dense and crystalline in case of long chain lengths. This is similar to the ordering of SAMs on Ag and Au substrates. SAMs of alkane thiols on Pd resulted in the enhancement of the resistance of the modified surface to corrosion which is independent of length of alkane chain. A complex palladium sulfide interphase was formed and revealed by XPS which resulted in the enhancement of SAM stability against corrosion. In addition, the reductive desorption potential of alkane thiols on Pd substrate is independent of the alkane chain length unlike Ag, Au, Ni and Pt. This is attributed to the absence of analogous metal–sulfide interphase in these cases [6]. Pd is very useful as a substrate for SAMs formation as a result of many physical characteristics. It has a very high affinity towards sulfur and it exhibits a smaller lattice (2.75 A°) compared to Au and Ag. Carlos R. Cabrera studied the modification of Pd substrate with l-cysteine. The SAM modified surface was characterized using cyclic voltammetry, electrochemical impedance spectroscopy, FTIR and reductive/oxidative desorption. XPS and FTIR revealed that l-cysteine molecules are adsorbed on the surface via S-atom leaving the carboxylic acid and amino groups free in the Pd surface for the biomolecules conjugation [4]. The electrochemical results proved that the electron transfer is enhanced and the charge transfer resistance decreased by SAM modification [4]. Furthermore, the SAMs based on alkanethiols on Pd show three fascinating features making them superior SAMs compared to those on other substrates of Au and Ag; i) SAMs can form etched structures with smaller edge roughness than the case of Au due to the variation in the grain sizes developing in thin films of metals that were deposited by evaporation onto silicon wafers, ii) fewer surface defects were obtained due to the higher selectivity against wet chemical etchants than those in case of Au and iii) they have potential impact in areas of sensors, catalysis and electronics [19]. A metastable Pd-S interphase is formed through the SAM formation of alkanethiols on Pd. The structure of the formed SAM is most similar to that formed in case of Ag. The Pd substrates don't oxidize easily and this is an obvious advantage of this case over Ag. SAMs of alkanethiols on Pd suffered from lower stability in air than the case of Au and Ag as the S species undergo oxidation within 2-5 days [19].

3.1.3.6. Zinc

Alkanethiols can form self-assembled monolayer on Zn substrate via chemisorption forming a highly hydrophobic well organized monolayer with few defects at atmospheric pressure. These SAMs oriented normal to Zn surface protecting the surface from oxidation in a neutral aqueous medium. They also provide a barrier against ions, oxygen and water diffusion therefore improve the corrosion resistance. Alkanethiols that have a second terminal functional group can act as a grafting agent so that they can bind to other layers chemically or electrochemically. Longer adsorption times results in the formation of increasingly organized SAMs. Longer alkyl chain length and longer adsorption time (48 hour) lead to higher hydrophobicity [15]. On the other hand, the nature of the chemical bond formed between the thiol sulfur and Zn was characterized using XPS. The thiol sulfur was bound to Zn(II) resulting in the formation of stable thiolates. It was proved that Zn(0)-SR was absent and Zn(II) thiolate was unusual stable. Zn substrate is oxidized to Zn(II) which then reacts with thiol sulfur forming an usual stable Zn(II) thiolate. It was observed that the thiol is unable to reduce Zn(II) contrary to other metallic species like Cu(II) which can be reduced by thiols [20].

3.1.3.7. Silicon

SAMs of organic molecules of different chain length can be well-organized on Si oxide or bare Si substrates using different strategies. They exhibit wide applications in molecular electronics. Fig. 3.5 shows the SAM over Si substrate for molecular electronic applications. The SAMs can be formed spontaneously through soaking the substrate of Si in SAMs precursor solution (like surfactant molecules $R(CH_2)_nSiX_3$ where X = Cl, OCH_3 or OC_2H_5 dissolved in alkane/carbon tetrachloride) or through other methods like vapor deposition. SAMs can be formed on Si oxide and bare Si substrate through; i) Silanization or vapor deposition, and ii) Si-C bond formation, respectively. A thin film of different metals (Al, Au, Hg, etc.) can be formed on the SAMs to construct molecular devices using them and measure their electrical properties. The self-assembled molecules contain three characteristic parts:

i) Head group like SiX_3: it has the ability to form chemical bond with surface atoms of substrate resulting in the grafting of the substrate with the surfactant molecules (exothermic step).

ii) Alkyl chain $(CH_2)_n$: The inter-chain van der Waals interactions help in the formation of well-ordered structure.

iii) Surface group (R): The terminal group that can be replaced with various functional groups for molecular electronics applications [21].

3.1.4. Assembly Process of Alkanethiols Over Gold

R. C. Salvarezza et al. studied the assembly process of alkanethiols over Au (111) substrates (Fig. 3.6). First, physisorption (Equation 3.1) occurs through van der Waals interactions resulting in disordered film as a gas-like state. After physisorption, thiol molecules are chemisorbed through the sulfur head group leading to the formation of a strong covalent bond of Au-S (Equation 3.2). This process takes place within some minutes. Thiol molecules (RS-H) lose their mercaptan H atom during the chemisorption step. The adsorption process of thiol molecules over Au (111) can be preceded through the following steps [9]:

$$CH_3(CH_2)_nSH + Au \longrightarrow (CH_3(CH_2)nSH)_{phys}Au \quad (\text{Physisorption})(3.1)$$

$$(CH3(CH2)nSH)physAu \longrightarrow CH3(CH2)nS-Au+1/2H2$$
$$(\text{Chemisorption}) \quad (3.2)$$

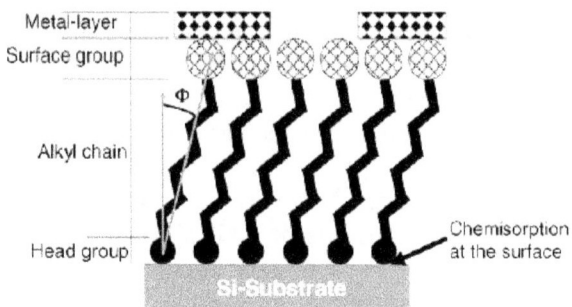

Fig. 3.5. A schematic diagram showing different parts of a self-assembled monolayer deposited on a Si-substrate suitable for molecular electronics. Φ is the tilt of the chain axis from the surface normal, reused with permission [21].

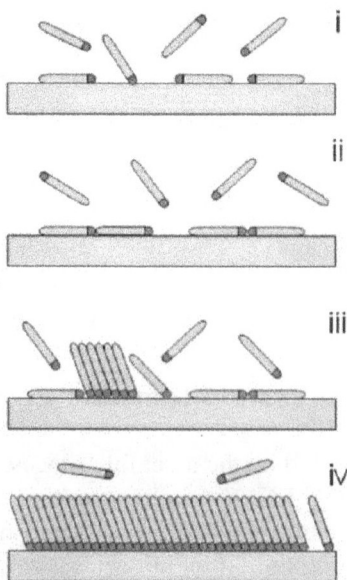

Fig. 3.6. Different steps taking place during the self-assembly of alkanethiol on Au(111): (i) physisorption, (ii) lying down phase formation, (iii) nucleation of the standing up phase, (iv) completion of the standing up phase, reused with permission [9].

3.2. Methods for SAMs Preparation

3.2.1. Chemical Deposition of SAMs

3.2.1.1. Solution Deposition

The most common method used for the SAM based thiol formation is the solution deposition or immersion growth. It includes the spontaneous adsorption of molecules from liquid solution to a solid substrate [22, 23]. SAMs are grown by the immersion of the substrate in the solution of the surface-active material. Chemical bond formation between the active material and substrate surface (mainly covalent bond between the substrate and the anchor site of SAM) and intermolecular interactions are the main leading forces in SAMs formation (Fig. 3.7) [22].

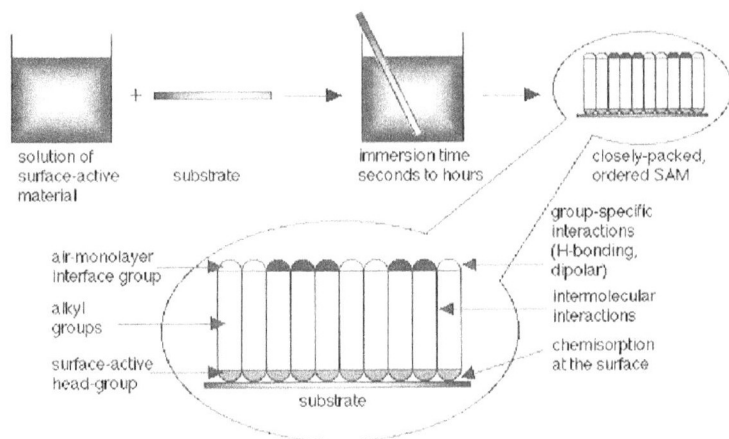

Fig. 3.7. Self-assembled monolayers are formed by simply immersing a substrate into a solution of the surface-active material. The driving force for the spontaneous formation of the 2D assembly includes chemical bond formation of molecules with the surface and intermolecular interactions, reused with permission [22].

SAM of alkane thiol molecules can be formed on a gold substrate in a well-ordered manner by simple immersion of Au substrate in alkane thiol solution. Alkane thiols with disulfides result in similar SAMs due to the S-S bond cleavage during adsorption. On the other hand, they show lower solubility than mono-S thiols and this may result in multilayer formation by precipitation. Organic solvents like methanol and ethanol are the most common solvents for SAM preparation. Hexane, toluene and benzene can be used in case of very long chains alkane thiols while water can be used for short chains alkane thiols. Low concentration (nanomolar) of alkane thiols solution and short times (seconds) are needed for low coverage SAMs or island formation. While, high concentration (micro to milli-molar) and long adsorption times (several hours or days) are required for densely packed and crystalline SAMs (high quality SAMs).

The SAM formation in case of Ag and Cu involves the reduction of Ag or Cu oxides into metal and alkane thiol molecules are oxidized into sulfonates which diffuse away from the surface as they are soluble. This is followed by the reaction of metal surface with the other alkane thiol in the solution forming the SAMs. The S binder in the thiol can't react with the oxide of Ag and Cu therefore it is necessary to reduce the oxide first. In case of GaAs substrate, the oxide film should be removed first by etching with NH_3. In case of Ni substrate, the native oxide film of NiO can be removed by electrochemical reduction in the presence of small amount of thiol in solution [23].

3.2.1.2. Gas Phase Deposition

Gas phase deposition is a famous method in which the evaporated alkanethiols adsorb onto the clean surface of metal or semiconductor usually in an ultra-high vacuum (UHV) chamber. For short alkanethiols, the UHV is a well-suited step as the short alkane thiols are highly volatile and can be easily evaporated under vacuum. This is applicable for all metal and semiconductor substrates. Gas phase deposition can be used for formation of highly ordered assemblies at sub-monolayer and full monolayer coverage through adsorption of alkanethiols and dialkyl disulfides (n < 10). Many precursors of SAMs lack the suitable vapor pressures and this is considered one important limitation of using gas phase deposition. On the other hand, working under controlled conditions is one important advantage of gas phase deposition as the substrate preparation, adsorption of molecules and characterization are achieved under very controlled conditions.

SAMs based on S can be easily formed on gold and silver substrates by immersion in solutions containing sulfide ions (S^{-2}, SH^- or SH_2) or by the exposure to gaseous species like S_2 or SO_2. The SAMs based on S can be formed on Ni substrate by immersion in solution of H_2S/HS^- prepared in acid medium. The preparation conditions (adsorption time, substrate type, concentration of solution, temperature, etc.) affect greatly the SAM structure and its defects in solution and gas phase deposition methods. The physical properties (ionic conductivity and electron transport) and redox behavior of SAMs in electrolytic solutions are affected by the presence of defects such as vacancies, domain boundaries and disordered chains. So that it is very crucial to draw a complete picture about the SAMs at the nano-scale level for future technological applications [23].

3.2.2. Electrochemical Deposition of SAMs

An alternative method for SAM formation is the electrochemical growth developed by Ying Zhuo et al. It is based on potentiodynamic control for formation of multilayer. The chemical formation of SAMs suffers from some drawbacks like possible competitive adsorption from other ions like supporting electrolyte. Also, the assemblies produced from chemical method may suffer from non-uniform film growth. The electrochemical growth is free from the as mentioned drawbacks. Also, the electrochemical method is selective and controllable method for SAM formation [24]. Atta et al compared the response of Au electrode modified with Cysteine SAM formed by chemical growth and electrochemical growth toward dopamine oxidation. The chemical growth involves soaking in 5 mmol L^{-1} cysteine/0.1 mol L^{-1} PBS/pH 2.58 for 5 minutes. While the electrochemical growth involves potential cycling at 10 mV/s within the limits of - 0.5 V to 1.0 V in a solution containing 5 mmol L^{-1} cysteine/0.1 mol L^{-1} KCl/1 mmol L^{-1} HCl for 30 minutes [24]. Better performance in terms of higher current response, less positive oxidation potential and higher reversibility were achieved at SAM prepared by chemical growth compared to that prepared by electrochemical growth (Fig. 3.8) [17].

3.3. Some Factors Related to the SAMs

3.3.1. Effect of Chain Length

The chain length of alkane thiol molecules have a great impact on the SAM formed over gold surface. It was reported previously that as the chain length increases, the SAM becomes more ordered. Also increasing the chain length results in the decrease of the charge transfer resistance of SAM modified electrode.

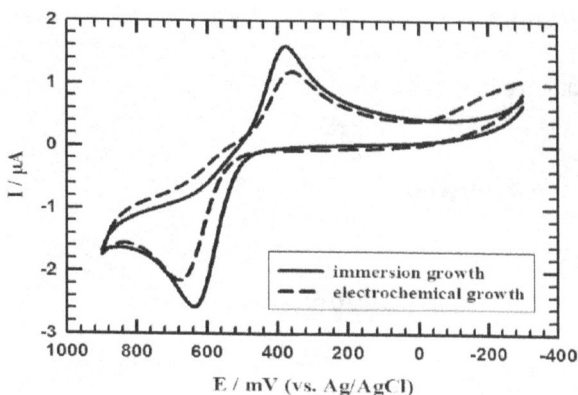

Fig. 3.8. CVs of 1 mmol L^{-1} DA/0.1 mol L^{-1} PBS/pH 2.58 at Au electrode modified with Cys SAM obtained by immersion (solid line) and electrochemical (dash line) growth method, scan rate 50 mV s^{-1}, reused with permission [17].

Also, longer chain length alkanethiol molecules needs shorter deposition time for highly ordered SAM formation. M. K. Sezgintürk et al fabricated an immunosensor based on 11-mercaptoundecanoic acid (MUA, HS(CH$_2$)$_{10}$CO$_2$H) SAM modified gold electrode. Also, they investigated the effect of using shorter chain length alkanethiol molecules like 3-mercaptopropionic acid (MPA, HSCH$_2$CH$_2$CO$_2$H). The results reported in their study are in a good agreement with the previously reported data. The deposition time needed for MPA SAM formation onto gold electrode is about 18 hours while that needed for MUA SAM formation onto gold electrode is 2 hours. The charge transfer resistance of MUA/gold is 4 kΩ which is much lower than that for MPA/gold (100 kΩ). MUA molecules create highly ordered and densely packed SAMs over the gold substrate due to the stronger van der Waals interactions between MUA chains compared to that for MPA [25].

3.3.2. Effect of Mixed SAMs

Homogenous single SAMs may suffer from some drawbacks that make the mixed SAMs more effective for some applications. The presence of unfavorable interactions between the neighboring end groups of the SAM may impede the attachment of drug to the SAM. A. Mirmohseni et al used mixed SAMs of 12-mercaptododecanoic acid and 1-nonanethiol on gold surface in the everolimus eluting coronary stents. 12-mercaptododecanoic acid is used as a functional thiolate and 1-nonanethiol is used as a diluting thiolate [26].

3.3.3. Effect of Number of S-atoms in the Thiol Compound

The number of sulfur atoms present in the organothiols molecules affects greatly the surface coverage of thiol molecules over the substrate surface. D.W.M. Arrigan et al studied the variation of surface coverage of l-cysteine and l-cystine over gold substrate with the corresponding concentration for each one. Fig. 3.9 shows the surface coverage variation with solution concentration for l-cysteine and l-cystine at gold electrodes. The electrochemical measurements are carried out in a solution free of l-cysteine and l-cystine therefore the data are non-equilibrium. The solid line in both cases represents the best fit non-linear least squares fit to Langmuir adsorption isotherm. In both cases, as the concentration of thiol solution increases the surface coverage of SAM over the surface increases till the saturation and formation of highly ordered and densely packed SAM. The calculated saturation surface coverage Γ_{sat} of l-cysteine is 2.7 ± 0.11 nmol cm^{-2} which is much higher than that in case of l-cystine (1.7 ± 0.17 nmol cm^{-2}).

The surface coverage of l-cystine over gold surface involves the coverage of l-cysteine monomer that results from the l-cystine dimmer. The adsorption of disulfides involves the cleavage of S-S bond and creation of adsorbed radicals of thiols (Equation 3.3):

$$RS\text{-}SR \longrightarrow 2RS^-_{(ads)} \qquad (3.3)$$

The adsorption of dialkyl disulfides onto gold results in the formation of a film similar to that from alkanethiols. The saturation surface coverage of l-cysteine monomers resulting from l-cystine is lower than that resulting from l-cysteine. This indicates that there is an observable difference in the molecular packing of molecules in both cases.

The adsorption of l-cystine (Equation 3.4) and its oxidative desorption (Equation 3.5) will proceed through the following equations, respectively:

$$RS\text{-}SR \longrightarrow 2RS^-(ads)\ (adsorption) \qquad (3.4)$$

$$RS^-_{(ads)} + 3\ H_2O \longrightarrow RSO3^- + 6H^+ + 5e^-\ (oxidative\ adsorption)\ (3.5)$$

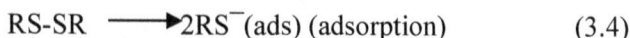

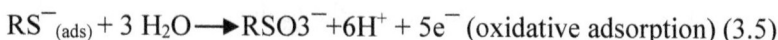

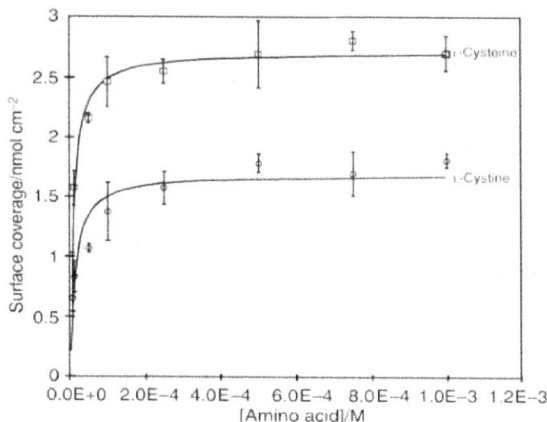

Fig. 3.9. Surface coverage dependence on solution concentration for l-cysteine and l-cystine at gold electrodes. The solid lines are best non-linear least-squares fits to the Langmuir isotherm, reused with permission [27].

In the present study, it is assumed that the oxidation of l-cystine involves five electrons. However Koryta and Pradac indicated that 10 electrons are released per l-cystine molecules. This means that not all S-S bonds are cleaved, they are partly dissociated. Therefore, a difference in the packing between dimeric l-cystine and monomeric l-cysteine over the gold surface is observed. Reynaud et al. proved that the oxidative desorption of l-cystine SAM from Au surface produced $RSSO_3^-$ confirming the incomplete dissociation of S-S bonds of l-cystine. The lower saturation surface coverage of l-cystine over gold surface than that of l-cysteine is attributed to the molecular packing of l-cystine in a more horizontal arrangement and the molecular packing of l-cysteine in a more vertical arrangement. Structural studies are needed to clearly understand the packing of l-cystine versus l-cysteine on the surface [27].

3.3.4. Effect of Time of Deposition in Chemical Growth Method

The surface coverage (θ, mol cm^{-2}) of SAM over the substrate is affected by immersion time of substrate in the SAM solution. The relation between the surface coverage and time can be described according to the following equation (Equation 3.6):

$$\theta(t) = [1-\exp(-kt)] \qquad (3.6)$$

In the equation, (k) is the adsorption rate constant. D. Merli et al studied the relation between (θ, nmol cm^{-2}) of SAM of 11-mercapto-undecylphosphonic acid (2.5 mM of MUPA) on gold surface with time (minutes). As the immersion time increases, the MUPA surface coverage increases till the formation of monolayer. As reported in literature, a complete adsorption of thiols on gold surface occurs after two hours. Merli et al prepared MUPA modified gold electrode by immersion in MUPA solution prolonged overnight in order to form highly rearranged and stabilized SAM [28].

3.3.5. Effect of Macro and Nano Au Substrate

There is a potential difference between the SAM formation on poly crystalline Au and nanoparticles gold. The experiment of desorption can give an evidence for alkanethiols SAMs formation on polycrystalline gold and gold nanoparticles modified electrode through S–Au bond. Atta et al studied the reductive desorption of Au-Cys and Au/Au$_{nano}$-Cys in 0.5 mol L^{-1} KOH (Inset of Fig. 3.10). Two cathodic peaks are observed at -736 mV and -1027 mV at Au-Cys electrode while they shift to -720 mV and -1008 mV at Au/Au$_{nano}$-Cys electrode. An obvious increase in the peak current of desorption step is observed upon modifying the bare Au electrode with gold nanoparticles. The surface coverage of cysteine SAM is calculated from the area under the desorption peak. It is 2.64×10^{-9} mol cm^{2} and 4.43×10^{-9} mol cm^{2} in case of Au-Cys and Au/Au$_{nano}$-Cys electrodes, respectively. Therefore, gold nanoparticles results in the enhancement of the immobilization amount of cysteine and the improvement of Au-S bond and stability of cysteine SAM [29].

Also, the oxidative desorption of (a) Au-Cysteine and (b) Au/Au$_{nano}$-Cysteine is achieved in 0.1 mol L^{-1} PBS/pH 2.58 (Fig. 3.10). It is anticipated that gold nanoparticles modification results in the increment of cysteine amount self-assembled on Au surface. Gold nanoparticles modification affects also the stability and structure of cysteine SAM [29].

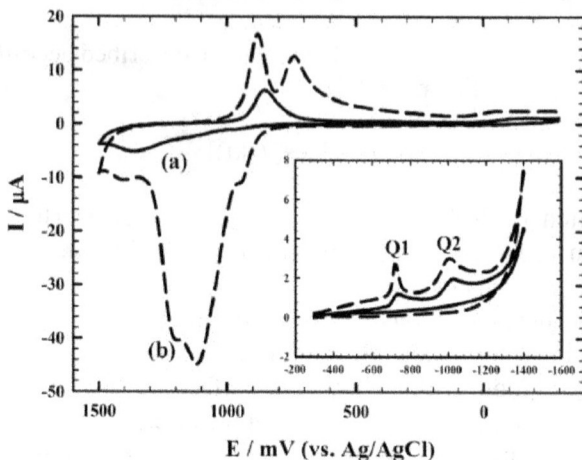

Fig. 3.10. CVs of the oxidative desorption of (a) Au-Cys and (b) Au/Au$_{nano}$-Cys in 0.1 mol L^{-1} PBS/pH 2.58. Inset; CVs of the reductive desorption of Au-Cys (solid line) and Au/Au$_{nano}$-Cys (dash line) in 0.5 mol L^{-1} KOH, scan rate 50 mV s^{-1}, reused with permission [29].

On the other hand, Atta et al studied the effect of repeated cycles stability up to 50 cycles of cysteine SAM modified gold electrode (Au-Cys) and gold nanoparticles modified gold electrode (Au-Au$_{nano}$-Cys). The electrochemical measurements were carried out in 1 mmol L^{-1} DA in 0.1 mol L^{-1} PBS/pH 2.58. As shown in Fig. 3.11(A), Au-Cys shows excellent stability via repeated cycles with the same current response for the 1st, 25th and 50th cycles. This means that a highly ordered and well-organized SAM of cysteine is formed spontaneously over the polycrystalline gold surface. Fig. 3.11(B) shows the comparison of 1st, 25th and 50th cycles at Au-Au$_{nano}$-Cys. The current response increases from 2.5 µA in the 1st cycle to 3.8 µA in the 25th and 50th cycles and the peak separation decreases from 120 mV in the 1st cycle to ~ zero mV in the 25th and 50th cycles. This indicates that reorganization of cysteine molecules Au-Au$_{nano}$ occurs by repeated cycles. This results in the enhancement of H-bond between Cys SAM and DA molecules leading to the enhanced DA diffusion and electron transfer rate. Cysteine molecules form a well-ordered and highly organized SAM over polycrystalline Au while they are well-reorganized over nanogold modified electrode via repeated cycles [17].

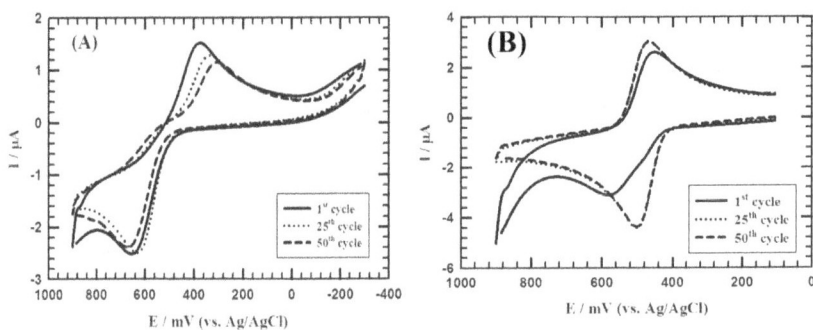

Fig. 3.11. The comparison of 1st, 25th, and 50th cycles of the repeated cycles stability of (A) Au-Cys and (B) Au-Au$_{nano}$-Cys electrodes in 1 mmol L^{-1} DA/0.1 mol L^{-1} PBS/pH 2.58, scan rate 50 mV s^{-1}, reused with permission [17].

3.4. Characterization of SAM Modified Electrodes

3.4.1. Electrochemical Characterization

3.4.1.1. EIS

Electrochemical impedance spectroscopy (EIS) is a perfect technique to monitor the properties of SAM. It can give information about the surface coverage, kinetics and mechanism of SAM formation process, holes/defects distribution and properties of the linked redox species. A decrease in the CV current of the $Fe(CN)_6^{4-}$ / $Fe(CN)_6^{3-}$ redox couple is observed upon modifying the bare gold electrode with a SAM of 3-mercaptopropionic acid (MPA). This may be due to the blocking effect caused by the SAM. From the Nyquist plot at the bare and SAM modified surface, we conclude that the modification of the electrode by a passive monolayer affects the charge transfer resistance, electrode capacitance and Warburg coefficient. An increase in the charge transfer resistance and Warburg coefficient and a decrease in the electrode capacitance take place upon modifying the Au with SAM of MPA. The EIS data agree with CV [30].

3.4.1.2. Electrochemical Desorption

Electrochemical desorption experiments (oxidative and reductive) can be used for further confirmation of SAMs of thiols on Au through S-Au bond formation.

3.4.1.2.1. Reductive Desorption

The reductive desorption of alkanethiols from gold surfaces is a one electron reduction process given by the following equation (Equation 3.7):

$$Au\text{-}SR + e^- \longrightarrow Au + R\text{-}S^- \qquad (3.7)$$

Reductive desorption can be used to confirm the formation of SAMs of 3-mercaptopropylphosphonic acid [HS-(CH$_2$)$_3$-PO$_3$H$_2$, MPPA] on Au substrate via S-Au bond. The reductive desorption step (reductive desorption of MPPA-Au in 0.5 M KOH, degassed solution) is a one electron reduction process involving the reductive desorption of the surface attached MPPA thiolates at about -1300 mV$\pm$10 mV. The surface coverage of MPPA on Au substrate can be determined from the charge consumed in the reductive desorption. The surface coverage is $5.6\pm0.3 \times 10^{-10}$ mol/cm^2 obtained from the current integration under the cathodic peak [31].

3.4.1.2.2. Oxidative Desorption

A monolayer of cysteine (RSH, where R is the amino acid functional groups) can be formed on the gold surface through the following reaction (Equation 3.8):

$$RS\text{-}H \longrightarrow H+ + RS^-(ads) + e^- \qquad (3.8)$$

The adsorbed l-cysteine on gold surface can be oxidized via oxidative desorption reaction through the following equation (Equation 3.9):

$$RS^-_{(ads)} + 3 H_2O \longrightarrow RSO_3^- + 6H^+ + 5e^- \qquad (3.9)$$

This reaction is accompanied by the formation of gold oxide over the gold surface. The cysteinesulfonic acid (or cysteic acid) is desorbed from the gold surface in the solution. Therefore, the SAM formation (one-electron spontaneous adsorption) and oxidative desorption processes (potentiodynamically driven) of l-cysteine on gold substrate involve six electrons including the oxidation of sulfur from -2 oxidation state to +4 oxidation state. The charge consumed in the oxidative desorption step can be used to calculate the surface coverage of l-cysteine SAM on the gold surface. Fig. 3.12 shows the oxidative desorption of l-cysteine SAM adsorbed on gold surface in aqueous solution of 0.01 M HClO$_4$ which is free from cysteine. Peak A in the cyclic voltammogram indicates the oxidation of l-cysteine SAM into RSO_3^- and formation of gold oxide film. While peak C is attributed to the reduction of the formed gold oxide film. Therefore, the differences in the charge values between peak A and C represents the charge needed for the l-cysteine SAM oxidation. Using Faraday's law one can calculate the number of moles of adsorbed l-cysteine on gold surface. The calculated surface coverage of adsorbed l-cysteine over the gold surface is 2.7 ± 0.11 nmol cm^{-2} [27].

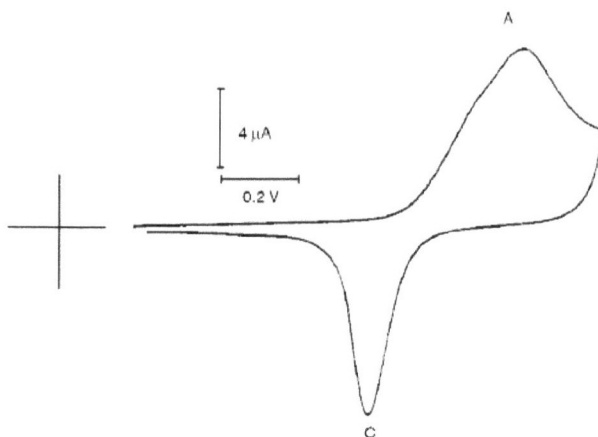

Fig. 3.12. Voltammogram of the oxidative desorption of l-cysteine from a gold electrode, following 5 min adsorption from 10^{-5} M l-cysteine in aqueous 0.1 M HClO$_4$ solution at open circuit. Voltammetry in 0.01 M HClO$_4$, sweep rate 50 mV/s, reused with permission [27].

3.4.2. Surface Characterization

3.4.2.1. XPS

XPS, X-ray photoemission spectroscopy, is a powerful tool that can be used to give a complete characterization of SAM. The presence of the desired elements in the film can be confirmed by XPS spectra. Also, the atomic composition of these elements can be evaluated. XPS can give information about the film thickness by evaluating the attenuation length of photoelectrons in the SAMs if the mean free path of the photoelectron in the studied layer is known. Also, it can give a characteristic image for the molecular packing of SAMs on the substrate surface. XPS can

117

distinguish between the perfect densely packed and imperfect SAMs on the substrate surface [32].

XPS can be used to identify the featured elements of Au electrode modified with 3-mercaptopropylphosphonic acid [HS-$(CH_2)_3$-PO_3H_2, MPPA]. The XPS spectra of Au/MPPA in the S(2p) and P(2p) regions are studied. The S(2p) is distinguished by two doublet peaks corresponding to the bound thiolates and unbound thiols. For bound thiolates, the $S2p_{3/2}$ and $S2p_{1/2}$ peaks appear at 162.0 eV and 163.2 eV, respectively. For unbound thiols, the $S2p_{3/2}$ and $S2p_{1/2}$ peaks appear at 163.6 eV and 164.8 eV, respectively. The ratio of the peak area of the bound and unbound sulfur species is 71:29 for the case of the study. There are two types of MPPA molecules on the gold surface similar to the case of carboxyl-terminated alkanethiols modified surfaces. In case of bound thiolates, the MPPA molecules can form a SAM directly attached to the gold surface through a strong S-Au bond. It is proposed that the sulfur atom of MPPA is bound directly to gold surface in bottom while the oxygen and phosphorus atoms of PO_3H_2 terminal group are located on the topmost layer of SAMs. While in case of unbound thiols, the MPPA molecules can attach to the gold surface through hydrogen bonding between phosphonic acid groups. The XPS spectra of P(2p) contain an asymmetric peak fitted as a spin doublet of $P(2p)_{3/2}$ and $P(2p)_{1/2}$ at 133.3 eV and 134.2 eV, respectively. The previous results confirm the self-assembling of PMMA on gold surface via S-Au bond with some unbound MPPA thiols [31].

3.4.2.2. FT-IR

FTIR can be used to confirm the formation of SAMs of alkanethiols on the Au substrate through Au-S covalent bond [33-35]. The FTIR spectra of the liquid mercaptopropionic acid (MPA) and that of Au-MPA are compared. The band at about 2560 cm^{-1} in case of MPA assigned to S–H stretching vibrations mode disappears in Au-MPA due to the formation of covalent bond between S atoms and Au substrate confirming the SAM formation on Au substrate.

3.4.2.3. Raman Spectroscopy

Raman spectroscopy can be used to identify the Au-S covalent bond in case of SAM of S-containing compounds on gold substrates via studying the various vibration modes. The Raman spectra for pure cysteine (Cys) and cysteine capped on gold nanoparticles (Cys/gold nanoparticles) are compared. The regions of interest are S-H and NH_3^+ bending and stretching modes. The spectrum of pure cysteine showed characteristic peaks of S-H stretching at 2546 cm^{-1}, C-H stretching at 2970 cm^{-1} and CH_2 asymmetric stretching at 3000 cm^{-1}. Also, the NH_3^+ bending vibrations (rocking and asymmetric) appear in the range of 1000-1600 cm^{-1} representing the zwitter ionic form of cysteine. From the spectrum of cysteine capped on gold nanoparticles, Raman absorption stretching band of S-H at 2546 cm^{-1} disappear due to the covalent bonding between S and Au due to the SAM formation. Also, there is a new band appears at 1050 cm^{-1} may be due to the combination of bending modes of NH_3^+ into a single mode [36].

3.4.2.4. SEM

Scanning electron microscopy can give a complete picture of SAM modification over the electrode surface. It gives detailed information about the morphological changes upon SAM formation and its relation with the electrocatalytic activity of the resulting SAM. The SEM of bare Au electrode and Au modified with SAM of cysteine are studied and analyzed. The bare Au electrode contains many defects and upon modification with cysteine SAM a smooth closely packed film is observed. A clear layer is obtained at SAM/Au compared to bare Au electrode. The previous result confirms the formation of cysteine SAM over AU surface [37].

3.4.2.5. TEM

Transmission electron microscopy can be used to give a characteristic image for SAM on polycrystalline gold and gold nanoparticles. The TEM of gold nanoparticles capped with cysteine and gold nanoparticles are compared. There is a distinguishable difference between TEM of both cases. In case of gold nanoparticles only a cluster of particles is observed while in case of gold nanoparticles capped with cysteine spherical nanoparticles with a corona of cysteine are observed. Also, the case of gold nanoparticles capped with cysteine appears as distinct and spherical entities while the pure gold nanoparticles appear as free colloids nanoparticles. In addition, TEM images confirm that gold nanoparticles are stabilized through the interaction with cysteine molecules [36].

3.4.2.6. AFM

Atomic force microscopy can be used as a good evidence of the formation of SAM over the surface of substrate and give information about the structure, molecules distribution and surface coverage. R. Oriňáková et al modified gold coated quartz crystal with a chromium under-layer (Au/Cr QCM crystal) with a SAM of 1-hexadecanethiol (1-HDT). Fig. 3.13 shows the AFM images of (a) pure Au/Cr QCM crystal and (b) 1-HDT SAM modified Au/Cr QCM crystal.

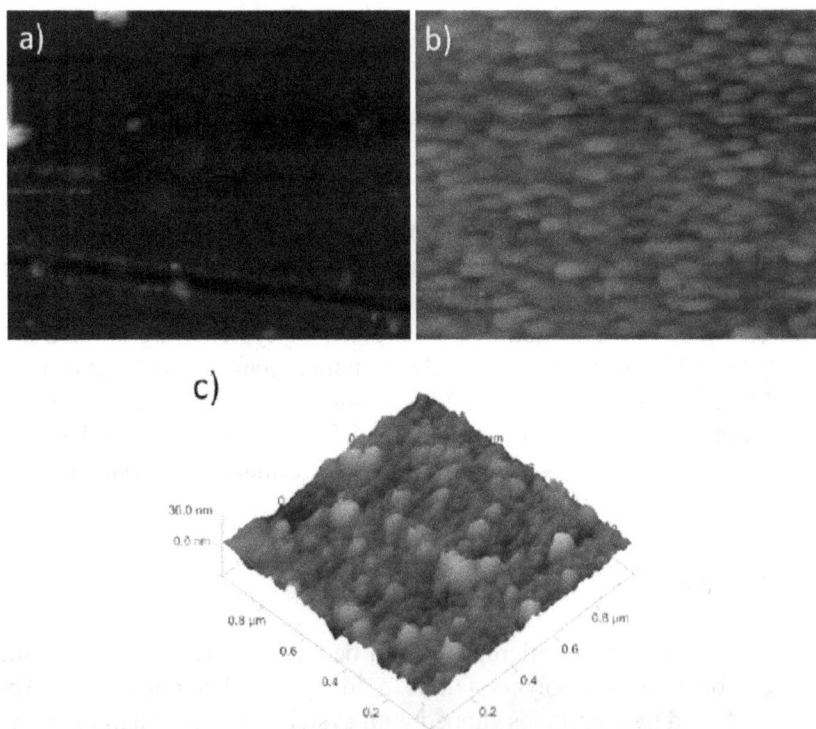

Fig. 3.13. AFM topography of a) pure Au/Cr QCM crystal surface, b) HDT covered Au/Cr QCM crystal surface and c) topography profile, reused with permission [38].

Fig. 3.13 (c) shows the topography profile of 1-HDT SAM modified Au/Cr QCM crystal. Fig. 3.13 (a) shows a clean smooth surface of Au/Cr

QCM crystal without any defects confirming the same probability for the adsorption of all thiol molecules all over the surface. While, Fig. 3.13 (b and c) shows well-ordered molecules, continuous layer and homogenous film of 1-HDT over the surface without defects [38].

3.4.2.7. NMR

The ^{1}H NMR can be used to confirm the interaction between Au substrate and alkanethiols like cysteine. The ^{1}H NMR spectra of pure cysteine and cysteine capped gold nanoparticles are compared. A quantitative shifting and splitting is observed in case of cys/Au as compared to the case of pure cysteine particularly in the proton region close to metal centers. Pure cysteine spectrum shows two peaks at δ 3.3 and δ 2.4 ppm representing the protons attached to α and β carbons, respectively. There is no observable shift in the peak at δ 3.3 ppm in spectrum in case of cys/Au. While the peak at δ 2.4 ppm is shifted to δ 2.6 ppm with quantitative splitting in cys/Au. This peak is due to the proton attached to β carbon which is very close to the metal center. This can create great in-homogeneity in the magnetic field and the local chemical environment. This shift is attributed to the interaction of sulfur of cysteine with gold nanoparticles. Also, the broadening of the peaks in cys/Au is attributed to the enhanced electron density and H-bond formation between the cysteine molecules capped gold nanoparticles [36].

3.4.2.8. UV-Vis

UV-Vis can be used to confirm the SAM formation over the surface of the substrate. The UV-Vis spectra of gold nanoparticles hydrosol and cysteine capped gold nanoparticles (Cys/Au) are compared. A strong absorption band is obtained in the gold nanoparticles spectra at 512 nm which is a distinguishable characteristic of plasmon resonance of gold. In the spectra of Cys/Au, a red shift is observed with band broadening indicating that there are some aggregations resulting from the surface modification of gold nanoparticles. Also, there is an observable color change from ruby-red (gold nanoparticles) to blue (Cys/Au) confirming the interaction of gold nanoparticles with cysteine through Au-S bond. Moreover, the interaction of gold nanoparticles with cysteine enhances the stability of the Cys/Au at room temperature compared to the case of the free gold nanoparticles. This is obvious from the spectra of Cys/Au taken after more than two months and that of free gold nanoparticles. The presence of cysteine stabilizes the gold nanoparticles through the formation of Au-S bond. Therefore, one can use UV-Vis to confirm that cysteine capped gold nanoparticles are more stable than the free gold nanoparticles [36].

3.4.2.9. Contact Angle Goniometry

SAMs modified electrodes can be characterized by determining the wettability through the measurement of water contact angle goniometry. Contact angle measurement can be used to elucidate the hydrophobic/hydrophilic nature of the surface. The SAM quality can be concluded from the measured wettability as the free energy of the proposed surfaces would affect the liquid drop shape [39, 40].

Nanqiang Li investigated the water contact angle for different SAMs (Penicillamine and Thiolactic acid) modified electrode. The advancing contact angle at gold electrode modified with Penicillamine (Pen) SAMs was 14°. The contact angle of Pen SAM is smaller than that of Thiolactic acid SAM (42°). Smaller contact angle values indicated surface with higher hydrophilic properties. Pen SAM contains two hydrophilic groups of COOH and NH_2 while Thiolactic acid SAM contains a hydrophilic COOH and hydrophobic CH_3 groups [39].

L.S. Koodlur measured the contact angle of bare gold and SAM of cobalt tetrasulfophthalocyanine (CoTSPc) modified gold electrode. The water contact angle decreased from 70±1° in case of bare gold to 43±2° in case of CoTSPc/Au. There is an obvious decrease in the contact angle value upon modification with SAM of CoTSPc due to the hydrophilic nature of sulfonic acid groups of the SAM resulting in hydrophilic surface [41].

3.5. Application of SAMs

As the surface of different substrates can be tailored with different types of SAMs in a unique and flexible methodology, the properties of the SAM modified surfaces are unique and perfect making them suitable for various applications. The SAMs are very attractive because of the large number of their potential applications. Several applications of alkanethiols SAMs in many fields of nanotechnology include electronics, spintronics, sensors, biosensors, catalysis, surface science, drug delivery and bio-recognition devices. Also, they have impact potential in fields like corrosion protection, lubrication and patterning of surfaces. In addition, highly ordered SAMs have potential impact in the preparation of metal–insulator–metal structures at the molecular level. Also, these assemblies can be applied as a backbone for resistors, insulators and capacitor devices in nano-electronics. In addition, the field of molecular electronics that depends on silicon will replace the silicon by Au-thiols bonds for electrical contact between the electrodes and the molecules [23]. Furthermore, SAMs modified electrodes have potential impact in electoanalytical applications as sensors and biosensors for the determination of different species ranging from metal ions to biomolecules. The applications of SAMs in sensing and biosensing will be our focus in this chapter. SAMs can be applied as sensors for neurotransmitters, drugs, metal ions, glucose and hydrogen peroxide. Also, they can be applied as DNA biosensor, immunosensor and cholesterol biosensor.

SAMs offer many advantages for sensors/biosensors which are summarized in the following points:

Ease of formation of highly ordered, stable and pinhole free monolayers.

Flexibility in the designing of SAM head group with different functional groups to tailor the surface properties.

Small amounts of biomolecules are needed to be immobilized on the SAM to form biomolecules monolayer.

Long term stability and good reproducibility offered by the proposed SAM.

SAM surface is similar to cellular membrane (microenvironment) making it suitable for the immobilization of biomolecules.

Resolving information about molecular level phenomena like hybridization of DNA, adsorption of proteins, etc. using surface techniques like atomic force and scanning tunneling microscopies [42].

3.5.1. Metal Sensors

SAMs modified electrodes offer highly stable, organized and well-ordered recognition system in the molecular scale [28, 43]. SAMs can change the properties of the proposed surface for the recognition of different analytes [44]. SAMs modified gold electrodes are widely used as metal ion sensors as they exhibit excellent sensitivity and selectivity [45]. Kubota prepared a sensitive sensor for Cu^{2+} by modification of gold electrode with a SAM of 3-mercaptopropionic acid. A reversible complex between Cu^{2+} and the carboxylic group of the SAM is described. A remarkable low detection limit of 1.8×10^{-14} mol l^{-1} is obtained [44]. Hibbert utilized l-cysteine modified gold electrode as a sensitive sensor for Cu^{2+} with very low detection limit below 5 ppb. The complexation of Cu^{2+} ions with cysteine is achieved through coordination with acidic –COOH and basic –NH_2 groups at pH 5.5 at low concentrations of Cu^{2+}. Each Cu^{2+} ion is connected with 2 cysteine molecules to form the preferred tetragonal geometry of complexed Cu^{2+} as the coordination number of Cu^{2+} is 4. At higher concentrations of Cu^{2+} (0.1 M), copper reacts with cysteine to form copper sulfide on the surface of the electrode [46]. Table 3.1 contains a summary of different SAMs modified electrodes as metal ions sensor.

Table 3.1. Different SAMs modified electrodes as metal ions sensor.

Electrode	Supporting electrolyte	Metal ions	Detection limit	Technique	Ref.
Screen printed electrode modified with gold nanoparticles then SAM of l-cysteine	0.04 M acetate buffer solution, pH 3.6	Al^{3+}	0.037 μM (8.90 ppb)	Differential pulse voltammetry	[45]
l-cysteine coated gold electrode	phosphate buffer solution, pH 7.54	Zn^{2+}	2.1 nM	Differential pulse voltammetry	[47]
gold electrode modified with SAMs of L-cysteine	phosphate buffer solution (pH 5.5)	Cu^{2+}	5 ppb	adsorptive stripping analysis	[46]
ω-mercaptoundecylphosphonic acid chemically modified gold electrode	0.1 M KCl, pH 3.5	UO_2^{2+}	0.3 μg L^{-1}	differential pulse adsorptive stripping voltammetry	[28]

3.5.2. Neurotransmitter Sensors

Neurotransmitters, endogenous chemicals, transmit signals from a neuron to a target cell through the synapse [17]. Dopamine (DA), an important neurotransmitter in the mammalian brain tissues, plays a very effective role in the renal, central, cardiovascular and hormonal systems functioning. Neurological disorders like Parkinson's and schizophrenia diseases are a result of dysfunction of the dopaminergic system in the central nervous system. From the disease diagnosis, the real time determination of DA amount is very necessary. Electrochemical methods have been used for the sensitive determination of DA with high sensitivity and easy miniaturization [31, 48, 49]. SAM of S-containing compounds can enhance the electrocatalytic activity toward dopamine oxidation, mediate the electron transfer rate and improve the sensing performance.

Zheng and Xia studied the electrocatalytic effect of 3-mercaptopropylphosphonic acid [$HS-(CH_2)_3-PO_3H_2$, MPPA] modified gold electrode toward the sensing of DA and AA. Electrochemical determination of DA is achieved at Au-MPPA in the presence of high concentrations of AA with good sensitivity and selectivity. This trend is attributed to the electrocatalytic activity of the proposed SAMs toward DA oxidation at physiological pH and the negligible response toward AA. MPPA contains negatively charged units which makes the MPPA an attracting interface for DA (positively charged) and repelling interface for AA (negatively charged). This allows the selective determination of DA without AA [31].

Galal et al introduced a novel sensor for DA based on SAM of cysteine on gold nanoparticles modified gold electrode in the presence of sodium dodecyl sulfate (SDS), Au-Au$_{nano}$-Cys-SDS. The electrostatic interaction between cationic DA and anionic SDS facilitates the DA diffusion through the positively charged SAM. In addition, the electrostatic interaction between positively charged SAMs of cysteine and anionic SDS enhances the reordering of cysteine molecules on gold nanoparticles. This results in the formation of hydrogen bond between DA and cysteine promoting faster electron transfer kinetics. The proposed sensor exhibits better stability in terms of repeated cycles stability and long term stability compared to Au-Au$_{nano}$ due to the strong Au-S bond. In addition, unusual high reversibility, almost zero or 15 mV peak separation, is achieved at the proposed sensor by repeated cycling confirming the role of Au-S bond in the enhancement of the electron transfer kinetics. Very low detection limit of 16 nmol L^{-1} is obtained for DA at the proposed sensor in the linear range of 30–100 µmol L^{-1}. Moreover, the simultaneous determination of DA, AA and paracetamol is achieved with good peak separation at the proposed sensor. Also, Galal et al compared the electrochemical behavior of cysteine SAMs on polycrystalline Au and Au nanoparticles for the first time. The cysteine effect is more pronounced at polycrystalline gold than that at gold nanoparticles. The effect of cysteine on gold nanoparticles is enhanced by SDS addition or repeated cycles allowing the reordering of cysteine molecules on Au_{nano} [16, 50]. Galal et al studied the long term stability of Au/Au_{nano} and $Au-Au_{nano}$-Cys modified electrodes toward the determination of dopamine up to one week (Fig. 3.14).

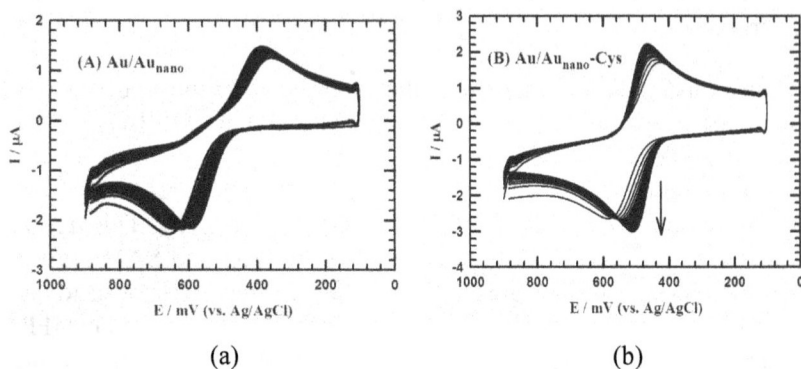

(a)   (b)

Fig. 3.14. Long term stability of: (a) Au/Au_{nano} and (b) Au/Au_{nano}-Cys after one week, 50 repeated cycles, 50 mV s^{-1} scan rate, reused with permission [17].

The current response of the 50th cycle at Au/Au_{nano} decreases by 26 % and 44 % and the peak separation increased to 120 mV and 180 mV after 3 days and one week of storage, respectively. On the other hand, the current response of the 50th cycle at $Au-Au_{nano}$-Cys decreases by 23.6 % and 35.6 % and the peak separation is zero and 15 mV after the same periods of storage, respectively. The previous results indicate that cysteine SAM on gold nanoparticles improve the reversibility and long term stability of $Au-Au_{nano}$-Cys due to formation of strong covalent Au-S bond [17, 50]. Table 3.2 summarizes the different SAMs modified surfaces for dopamine sensing.

Epinephrine (EP) is an important neurotransmitter that exists in the mammalian hormonal and central nervous system. It affects the blood pressure regulation and the heart rate. Low levels of EP are found in patients Parkinson's disease [29]. SAMs of thiols proved their electrocatalytic action toward the oxidation of epinephrine.

Table 3.2. Different SAMs modified electrodes as dopamine sensor.

Electrode	Supporting electrolyte	Detection limit	Sensitivity	Ref.
N,Ń-[1,10-Dithiobis(phenyl)] bis(salicylaldimine) (DTPS) self-assembled monolayer on Au electrode	pH 6.0, 0.1 mol L^{-1} phosphate buffer solution (PBS)	3.0×10^{-8} mol L^{-1} in the presence of 1 mmol L^{-1} AA	Not reported	[49]
Bean sprout peroxidase – cysteine SAM–Au electrode	pH 6.0, 0.1 mol L^{-1} PBS	4.78×10^{-7} in the presence of 6.0×10^{-5} mol L^{-1} hydrogen peroxide	1.65×10^{5} μA/mol L^{-1}	[51]
Au nanoarray electrode modified with L-cysteine SAM and gold colloid	PBS (pH 7.4)	5.0×10^{-10} mol/L		[52]
three-dimensional gold nanodendrite modified with SAM of 8-mercaptooctanoic acid	0.1 M tris buffer solution, pH 7.4	20 nM	1.43 ± 0.02 nA/mM	[53]
Cysteamine SAM modified Au electrode	0.1 M PBS (pH 2.0), 0.1 M NaClO$_4$	2.31 μM in the presence of 1.0 mM AA.	0.014 μA μM^{-1}	[54]
N-acetylcysteine modified Au electrode	0.1 M PBS, pH 6.5 containing 0.1 mol l^{-1} KCl	8.0×10^{-7} mol/L	0.01712 μA μM^{-1}	[55]
Thiolactic acid self-assembled gold electrode	0.05 M PBS, pH 6.81	3 μM	1753.3 μA M^{-1}	[56]
Homocysteine monolayer modified Au electrode	0.1 M HAc-NaAc buffer, pH 4.3	5.0×10^{-7} mol/L	0.7957 nA μM^{-1}	[57]
3,3`-dithiodipropionic acid self-assembled monolayers	0.1 mol L^{-1} phosphate buffer at pH 7.0	Not reported	0.098 A/mol L^{-1}	[58]
Penicillamine ((CH$_3$)$_2$C(SH)CH(NH$_2$)COOH) self-assembled gold electrode	0.1 M phosphate buffer, pH 6.81	4 μM	0.134×10^{4} μA M^{-1}	[39]
Au-cystamine-gold nanoparticles	0.1 M PBS, pH 7.2	0.22 μM in the presence of 0.1 mM AA	Not reported	[60]
Platinum Microelectrode/Au-NPs/3-mercaptopropionic acid SAM		7 nM	0.341 nA μM^{-1}	[61]

Galal et al investigated the electrochemistry of EP at bare Au electrode and cysteine SAM modified Au electrode. The anodic peak current of

EP increased and the anodic peak potential shifted to less positive potential upon modifying the Au electrode with cysteine SAM resulting in enhanced electron transfer kinetics. Cys-Au affects the mechanistic reaction of EP. There are two types of interactions as shown in Fig. 3.15: substrate-molecule and molecule-molecule interactions. The substrate-molecule interaction resembles the interaction of Au and cysteine via Au-S bond. The molecule-molecule interaction occurs between the cysteine and EP molecules through hydrogen bond formation between the nitrogen of cysteine SAM and hydrogen of the hydroxyl phenol of EP. Therefore, the hydroxyl-phenol on the benzene ring is more activated, the bond between hydrogen and oxygen becomes weak and the electron transfer becomes easier through N/H–O bond [29].

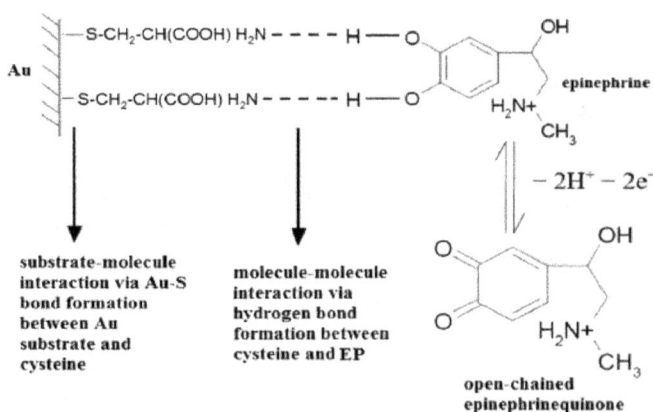

Fig. 3.15. Schematic model of the electrocatalytic oxidation of EP at Au-Cys electrode via substrate-molecule and molecule-molecule interactions, reused with permission [29].

In addition, Galal et al studied the electrochemistry of EP at cysteine SAM on gold nanoparticles modified Au electrode in the presence of sodium dodecyl sulfate, Au/Au$_{nano}$-Cys/SDS. Gold nanoparticles modification increase the cysteine amount immobilized on the surface and enhance the stability of cysteine SAM and Au-S bond. The proposed sensor can detect EP in presence of excess amount of uric acid and glucose. Also, it can simultaneously determine EP in the presence of AA and paracetamol with good potential peak separation. Very low detection limit of 0.294 nM is obtained for EP at the proposed sensor [29, 50]. Table 3.3 summarizes the different SAMs modified surfaces for epinephrine sensing.

Table 3.3. Different SAMs modified electrodes as epinephrine sensor.

Electrode	Supporting electrolyte	Detection limit	Sensitivity	Ref.
Au/mercaptopropionic acid (MPA) Au/MPA/cysteamine(CA)/gold naoparticles (Au-NPs) Au/Au-NPs/MPA/CA/Au-NPs	60 mM phosphate buffer (pH 7)	0.042 µM 0.053 µM 0.04 µM	Not reported	[33]
Gold electrode/gold nanoparticles (Au-NPs)/ meso-2,3-dimercaptosuccinic acid/Cysteamine/Au-NPs layers	phosphate buffer (60 mmol dm^{-3}, pH = 7)	2.8×10^{-5} mmol dm^{-3}	7 A cm^{-2} mol^{-1} dm^3	[61]
L-cysteine SAMs/gold electrode	0.5 mol l^{-1} phosphate buffer solution of pH 7.0	1.0×10^{-8} mol l^{-1}	0.7125 µA/ µM	[34]

Norepinephrine (NE) is an organic cation secreted by the adrenal medulla and exists in the nervous tissue and biological fluids. It has a great effect on the muscle and tissue control. Also, it decreases peripheral circulation, excites the arteriole contraction and initiates lipolysis in adipose tissue. Therefore, it is very important to determine NE quantitatively. Gold electrode modified with SAM of 2-Mercaptoethanol (ME/Au SAMs) has been used as an electrochemical sensor for NE. ME/Au SAMs shows higher current response, less positive oxidation potential and smaller potential peak separation compared to bare Au electrode toward NE oxidation. Very low detection limit of 7.0×10^{-7} mol /L is obtained and the relative standard deviation is 1.2 % for NE solution $(1.0\times10^{-5}$ M, n=11) [35]. Also, Wang et al utilized L-cysteine SAMs/gold electrode as norepinephrine sensor. The SAM accelerates the redox behavior of norepinephrine resulting in very low detection limit of 8.0×10^{-8} mol l^{-1} [34]. These achievements are mainly attributed to the hydrogen bond formed between the SAM and the studied analyte.

3.5.3. Sensor for Pharmaceutically Important Drugs

SAMs have potential applications in the interfacial chemistry and life science. They show identical features as that of biomembranes in the function and configuration. Also, they show great long term stability [62, 63]. Therefore, SAMs modified electrodes have been utilized in the sensitive determination of many pharmaceutically important drugs.

Isoproterenol is a pharmaceutically important drug used for cardiac chock, heart attack, bronchitis, primary pulmonary hypertension and allergic emergencies. Mazloum-Ardakani constructed an electrochemical sensor for isoproterenol based on the modification of glassy carbon electrode with a layer of gold nanoparticles then further modified with a layer of 2-(2,3-dihydroxyphenyl) benzothiazole (GCE/AuNPs/DPB). The SAM of DPB has an important role in the electrocatalysis of isoproterenol oxidation through electrochemical-chemical mechanism. The proposed sensor can determine isoproterenol in the presence of uric acid simultaneously. A very low detection limit of 82 nM is obtained [63].

On the other hand, ferulic acid is an anti-oxidative, anti-inflammatory, analgesic and antithrombotic agent. It is an active medicine that has wide applications in the treatment of coronary heart and cerebrovascular diseases. It plays an important role in the fields of health food, cosmetic and nutrition restorative. Gold electrode chemically modified with l-cysteine SAM is utilized for the electrochemical determination of ferulic acid (L-Cys/SAM-Au/CME). Upon modifying the Au electrode with cysteine SAM, the oxidation peak current increases obviously and the electron transfer rate is enhanced indicating the electrocatalytic effect offered by SAM. The electrocatalytic oxidation of ferulic acid at the proposed sensor is achieved via hydrogen bond formation between the ferulic acid and l-cysteine which facilitates the electron transfer through N...H-O. A very low detection limit of 1.2×10^{-7} M is obtained [62].

3.5.4. Sensor for Ascorbic and Uric Acids

Ascorbic acid (AA) is considered as anti-oxidant and its common name is vitamin C. It is used for the complement of the inconvenient dietary capacity. It is commonly used for mental illness, common cold, infertility and some clinical aspects of HIV infections. Uric acid (UA) is coexisting with AA in biological fluids like blood and urine. UA is the primary end product of purine metabolism. Its abnormal level is a mark of some diseases like Lesch-Nyhan, gout and hyperpiesia disease [64]. SAMs modified electrodes offer high sensitivity, good selectivity, enhanced conductivity, fast electron transfer kinetics and long term stability. Wang et al used l-cysteine SAM modified gold electrode (Cys/Au) as an electrochemical sensor for AA and UA. The SAM modified electrode shows excellent response toward the oxidation of AA and UA compared to the response at bare gold electrode. Moreover, the SAM modified surface can separate the overlapped peak of AA and UA in their mixture with a good potential peak separation of about 236 mV. Two well-defined oxidation peaks (which are overlapped at the bare electrode) are obtained at 26 and 262 mV for AA and UA, respectively at the SAM surface. The modified surface exhibits good selectivity, high sensitivity, excellent performance and antifouling characteristics. Very low detection limit of 2×10^{-6} M and 1×10^{-5} M are obtained for AA and UA, respectively [65]. Table 3.4 summarizes different SAMs modified electrodes used as ascorbic and uric acids sensor.

Table 3.4. Different SAMs modified electrodes as ascorbic and uric acids sensor.

Electrode	Compound of interest	Supporting electrolyte	Detection limit	Sensitivity	Ref.
Supramolecular film containing ferrocene on l-cysteine SAM modified gold Electrode	AA	0.1 mol l^{-1} BR buffer solution, pH 5.8	2.0×10^{-7} mol l^{-1}	0.0087 $\mu A/\mu M$	[37]
Self-assembling gold nanoparticles to the surface of the l-cysteine-modified glassy carbon electrode	AA UA	0.10 mol L^{-1} PBS (pH 7.0)	3.0×10^{-6} mol L^{-1} 2.0×10^{-7} mol L^{-1}	21.69 $\mu A/mM$ 0.769 $\mu A/\mu M$	[64]
Gold nanoparticles immobilized on 1,6-hexanedithiol modified gold electrode	AA	0.2 M phosphate buffer solution, pH 7.2	1 μM	Not reported	[66]
Positively charged gold nanoparticle and l-cysteine film on an Au electrode	AA	0.1 M PBS (pH 6.5)	1.5×10^{-7} M	Not reported	[67]
Gold electrode modified with a SAM of mercaptobenzimidazole	UA	0.1 M phosphate buffer solution (pH 7.2)	1 μM	0.0149 $\mu A/\mu M$	[8]

3.5.5. Immunosensor

Immunosensors are used to detect disease related substances which known as biomarkers in clinical diagnostics. Antibodies immobilized on

131

the biosensor surface are used to capture specific biomarkers [68]. The detection process of immunosensors is being related to changes in mass, resistance or capacitance at the sensor surface. Therefore, there are different types of immunosensors like impedimetric, piezoelectric, amperometric, capacitive and etc. Functionalized platforms for the immobilization of different antibodies and enzymes can be constructed with the aid of self-assembled monolayers [69]. SAMs offer characteristic features for biomolecules immobilization like simple strategy for SAMs formation, appropriate surface for the immobilization of biomolecules, small amount of biomolecules is required to be immobilized on the SAM modified surface and flexibility in designing the SAM head group with different functional groups [70, 71].

R. Yuan et al constructed amperometric immunosensor based on embedded antibodies on gold nanoparticles with nanocomposite SAM of SiO_2/thionine. First, a SAM of l-cysteine was self-assembled on gold electrode then two layers of gold nanoparticles were created using a linker of SiO_2/thionine. Finally, carcinoembryonic antibody (anti-CEA) which represents immuno-reagent model is immobilized on the surface. Under the optimum conditions, the proposed sensor is very sensitive to carcinoembryonic antigen (CEA) in the linear response of 1.00 and 100.00 ng/mL and detection limit of 0.34 ng/mL. Also, applicability for the proposed sensor to detect CEA in blood serum achieved with good results [24]. Different SAMs modified electrodes as immunosensors are summarized in Table 3.5.

3.5.6. DNA Sensor

DNA sensors have received much interest in the field of gene analysis, genetic disorders identification, tissue matching and forensic application. Electrochemical DNA biosensors are very promising for DNA detection and diagnosis due to many advantages such as low cost, sensitivity and rapid response. SAMs modified gold electrodes have been utilized as a building block in DNA sensors due to their characteristic features. SAMs have been utilized for the covalent immobilization of DNA on the surface of study [73]. DNA sensors based on alkylthiols SAMs exhibit high sensitivity, good selectivity and unexpected long term stability. SAMs can form very strong bonds with the gold substrate via Au-S bond and with the outermost layer through different attachment points. This would result in prolonged storage stability for DNA sensors based on alkanethiols SAMs [74]. In addition, SAMs based DNA sensors offer low overpotential, fast response time and antifouling characteristics [75].

Table 3.5. Different SAMs modified electrodes as immunosensors.

Electrode	Immunosensor type	For the detection of	Linear response	Ref.
SAM of 11-mercaptoundecanoic acid modified gold electrode	Impedimetric	vascular endothelial growth factor (VEGF)	1 to 6 ng/mL human VEGF	[25]
SAM of 16-mercaptohexadecanoic acid (MHDA) on an AT-cut quartz crystal's Au electrode surface	Piezoelectric	Escherichia coli O157:H7 (one of the most dangerous pathogens)	103–108 colony-forming units CFU/ml	[72]
antibody-embedded gold nanoparticles and SiO$_2$/Thionine self-assembled layers	Amperometric	carcinoembryonic antigen (CEA)	1.00 and 100.00 ng/mL with a detection limit of 0.34 ng/mL	[24]
cysteamine SAMs modified gold nanoparticles	Impedimetric	human epidermal growth factor receptor-3 (HER-3)	0.2 to 1.4 pg mL^{-1}	[70]
self-assembled monolayers and tyramine electropolymerization over the gold surface electrode	Capacitive	aflatoxin B1	3.2×10^{-13} M to 3.2×10^{-7} M	[68]
self-assembled thioctic acid onto the gold surface	Enzymatic	fibrinogen	Not reported	[71]
mixed mercaptohexadecanoic acid and 1,2-dipalmitoyl-snglycero-3-phosphoethanolamine-N-caproyl biotinyl SAMs on gold	Capacitive	haemoglobin	10^{-8} to 10^{-6} M	[69]

Jinghong et al constructed DNA biosensor based on the immobilization of target DNA over gold nanoparticles modified cysteamine SAM on gold electrode. Gold nanoparticles modified SAM of cysteamine result in the enhancement of electrode surface area and the immobilized amount of single stranded DNA (ssDNA). Then the ssDNA immobilized on Au electrode is hybridized with silver nanoparticles/oligonucleotide DNA probe. This is followed by the release of silver atoms established on the DNA hybrids in the solution by oxidative metal dissolution. The indirect determination of the released Ag^I ions can be achieved at a carbon fiber microelectrode via anodic stripping voltammetry. The linear range of the proposed DNA sensor is 10-800 pmol/l with a very low detection limit of 5 pmol/l [76]. Table 3.6 gives a brief summary of different SAMs modified electrodes as DNA biosensors.

Table 3.6. Different SAMs modified electrodes as DNA biosensors.

Electrode	For the detection of	Linear range	Detection limit	Ref.
Gold colloid particles associated with a cysteamine monolayer on gold electrode surface with silver nanoparticle label	DNA	10-/800 pmol/l	5 pmol/l	[76]
Monolayers of cysteine linked peptide nucleic acid (PNA) assembled on gold electrodes	DNA	Not reported	Not reported	[77]
Ternary thiolated self-assembled monolayers of a thiol-derivatized specific ss-oligonucleotide capture probe, 1,6-hexanedithiol and 6-mercapto- 1-hexanol	DNA	Not reported	Not reported	[74]
SAM of 1,8,15,22-tetraaminophthalocyanato-nickel(II) on glassy carbon electrode	Guanine "building blocks of both DNA and RNA"	10 to 100 μM	3×10^{-8} M	[78]
Thiolated single stranded deoxyribonucleic acid probe (ssDNA-SH) onto gold (Au) coated glass electrode	Vibrio cholerae	100-500 ng/μl with sensitivity of 0.027 μA/ng cm^{-2}	100 ng/μl	[73]
Gold mercaptopropionic acid self-assembled monolayers (Au-MPA SAMs) functionalized with 5-amino-1,10-phenanthroline-Fe(II) complex (5Aphen-Fe(II)) through covalent binding	Guanine	1.0 to 100.0 μM	0.17 μM	[75]

3.5.7. Hydrogen Peroxide Sensors

Hydrogen peroxide has a potential significance in different fields like pharmaceutical, environmental, clinical and food analysis. It is a marking molecule in the regulation of various biological processes like vascular remodeling, activation of immune cell, apoptosis, closure of stomata and the growth of the root. In addition, the determination of hydrogen peroxide is very important in the detection of some biologically important materials like glucose, cholesterol, etc. [41, 79-81].

Different modified electrodes have been fabricated including SAMs for the sensitive detection of H_2O_2. SAMs can be used for the immobilization of various types of proteins and enzymes. Also, they can be utilized as a cross linker between two similar or different layers in order to enhance the sensitivity of the proposed sensor [79, 80]. L-cysteine SAM on gold nanoparticles modified electrode exhibits lower background current, faster electron transfer rate and higher electrocatalytic activity. The presence of l-cysteine SAM on gold nanoparticles enhances the conductive area compared to gold nanoparticles alone [81]. In addition, certain types of SAM (cobalt tetrasulfophthalocyanine) on gold electrodes may act as electron antennae that would promote the electronic communications between the gold substrate and the solution at the interface [41].

Ruo Yuan et al constructed a H_2O_2 biosensor based on a hemoglobin (Hb)/gold colloid (nano-Au)/l-cysteine(l-cys)/nano-Au/Pt nanoparticles (nano-Pt)–chitosan(CHIT) composite film-modified platinum disk electrode. Firstly, Pt–CHIT inorganic–organic nanocomposite containing positively charged –NH_2 group are formed on the platinum disk. Then, based on the electrostatic interactions between oppositely charged species, negatively charged gold nanoparticles are immobilized on the Pt–CHIT nanocomposite. Then, based on covalent bonding between S and Au, self-assembly monolayer of l-cysteine is formed on the Au surface. After that a second layer of gold nanoparticles is immobilized on l-cysteine SAM through the –NH_2 group of l-cysteine in order to enhance the conduction characteristics of the sensor and the electron transfer rate. Finally, hemoglobin is adsorbed on gold nanoparticles due to its selectivity toward H_2O_2 detection. The presence of l-cysteine SAM promotes the electron transfer rate as confirmed by the decreased charge transfer resistance [79]. The proposed sensor exhibits good linear range of $1.4 \times 10^{-7} - 6.6 \times 10^{-3}$ M, low detection limit of 4.5×10^{-8} M, high sensitivity of 17.62 μA/mM within 10 s, good selectivity and enhanced stability. Table 3.7 contains a brief summary of different SAMs modified electrodes as H_2O_2 sensors with their linear range, sensitivity and detection limit.

Table 3.7. Different SAMs modified electrodes as H_2O_2 sensors.

Electrode	Linear range	Sensitivity	Detection limit	Ref.
Hemoglobin on gold colloid/ l-cysteine/ gold colloid/ nanoparticles Pt–chitosan composite film-modified platinum disk electrode	1.4×10^{-7} - 6.6×10^{-3} M	17.62 μA/mM	4.5×10^{-8} mol/L	[79]
Horseradish peroxidase immobilized on self-assembling gold nanoparticles to a thiol-containing sol-gel network ((3-mercaptopropyl)-trimethoxysilane sol-gel solution)	5 μM - 10 mM	Not reported	2.0 μM	[80]
Horseradish peroxidase/l-cysteine–goldparticle nanocomposite/nafion memebrane/modified glassy carbon electrode	1.60×10^{-5} - 1.10×10^{-3} M	Not reported	5.50×10^{-6}	[81]
SAM of a water-soluble cobalt tetrasulfo-phthalocyanine on a gold substrate	1-20 μM	Not reported	0.4 μM	[41]

3.5.8. Glucose Sensors

Determination of glucose is very crucial as diabetes mellitus becomes one of the major worldwide diseases. Diabetes results in intense deterioration in vision, heart and kidney. Diet control and/or insulin injections are the strategies that the diabetic patients follow to survive with their disease. Therefore, the permanent determination of glucose scale becomes urgent and critical. Glucose sensors are very necessary in clinical applications as the glucose level is a crucial marker of diabetes. Enzymatic and non-enzymatic glucose sensors have been utilized to monitor glucose grade with high sensitivity and selectivity. Self-assembled monolayers proved their potential role in the design of both glucose sensors, enzymatic and non-enzymatic. The conducting

characteristics of the SAMs and its ability to pass the electrical signals
through it will affect the sensor efficiency [82, 83].

In enzymatic glucose sensing, SAMs can be used to modify the gold
surface to boost its binding affinity with glucose oxidase enzyme. Thus,
the enzyme can be chemically bound to the gold surface through the
SAM [82, 83]. Patel et al employed SAMs of alkylthiols like 11-amino-
1-undecanethiol hydrochloride, 1-hexadecanethiol (1-hexadecanthiol)
and 1,9-nonanedithiol to modify the gold surface to enhance the adhesion
of glucose oxidase enzyme (GOx). Then, GOx is immobilized over the
modified gold surface. The presence of the SAMs over the gold enhances
the long term stability of the enzymatic sensors compared to the case of
bare gold electrode due to the strong adhesion of the GOx on the surface.
The unmodified gold electrode suffers from very short term stability due
to the dissolution of the GOx immobilized on the surface in the test
solution [82].

H. Ohnuki et al proved the role of SAM in the modification of gold
electrode for further modification with GOx and investigated the effect
of different types of SAMs. He employed SAMs of terminal COOH
groups (3-Mercaptopropionic acid (MPA), 11-mercaptoundecanoic acid
(MUA) and 4-mercaptobenzoic acid (MBA)) to modify the gold surface.
Then, GOx was covalently immobilized over SAM/Au and then nano-
clusters of Prussian Blue (PB) in the form of Langmuir–Blodgett films
were formed on the GOx surface. PB is known with its mediating ability
for glucose determination. The SAM type can affect the amount of the
immobilized GOx and further affect the sensor sensitivity. To study the
effect of the SAM type on the immobilized amount of GOx, the proposed
surface morphology of Au/SAM/GOx (SAM with many defects and
closely packed SAM). The GOx clusters exhibited large surface area
compared to SAM molecules size. Therefore, the defects and the
pinholes of SAM wouldn't affect the immobilized amount of GOx. Also,
there are many chemical bonds that will be available to immobilize a
single crystal of GOx. The amount of GOx immobilized on the different
SAMs was the same therefore it is necessary to confirm that the sensor
sensitivity is affected only with the SAM type.

The amperometric response of the proposed sensor, Au/SAM/GOx/PB,
toward glucose determination was studied using the different SAMs
(MPA, MUA and MBA). Au/MUA/GOx/PB showed very small
amperometric response and Au/MPA/GOx/PB showed slightly

increased response with glucose injection and the response time for both of them is about 20 s. While, Au/MBA/GOx/PB showed enhanced amperometric response and the response time was about 3 s. The current density of the different SAMs can be arranged in the following order; MBA>MPA>MUA with the ratio of 13.9:6.0:1.0, respectively. In addition, higher sensitivity of 50 nA/ cm^2 mM, wider linear range of 12.5 μM-70 mM and lower detection limit of 12.5 μM were obtained at Au/MBA/GOx/PB. These results confirmed the high conductivity of MBA SAM which affects the sensitivity and performance. Also, higher reproducibility was obtained at MBA as the relative standard deviation for 11 measurements was 2.1 %, 4.0 % and 11.7 % at MBA, MPA and MUA SAM based sensors, respectively. In addition, the long term stability was studied for 17 days in 1 mM glucose/pH 7.0. Longer term stability was obtained at MBA based sensor for 13-14 days compared to the other SAMs based sensor which were stable up to only one week. The ultra-thin MBA SAM showed a high conductivity nature and further high electron transfer rate resulting in the highest sensitivity, shortest response time and the most stable response. MBA confirms its role as a linker to bind GOx on the gold electrode as it forms a SAM with terminal COOH group which can be bound chemically with diversity of functional moieties. MBA has an aromatic moiety results in the formation of highly conductive SAMs with high electron transfer rate [83].

On the other hand, the SAM plays an essential role in the design of non-enzymatic glucose sensors resulting in enhanced sensitivity and improved electron transfer kinetics. Shengshui Hu et al fabricated non-enzymatic glucose sensor based on the electrodeposition of Cu particles on SAM of hexanethiol modified gold electrode, Cu/SAMs/Au. SAM contains pinholes acting as an array of microelectrodes enhancing the electron transfer rate. Upon electrodeposition of Cu, Cu microparticles can settle in the pinholes of the SAM resulting in highly mechanically stable SAM containing Cu particles. Cu/SAM/Au showed higher sensitivity toward glucose oxidation in alkaline medium and better selectivity toward glucose in presence of different interferents like ascorbic and uric acids. A linear range of 3.0 μM to 10 mM, sensitivity of 0.1116 A/M and detection limit of 0.7 μM were obtained at the Cu/SAM modified surface compared to Cu/Au electrode [84]. The better performance of Cu/SAM/Au compared to Cu/Au toward glucose oxidation can be discussed in terms of scanning electron microscopy. An obvious difference in Cu sedimentation is observed between the two cases of Cu/SAM and Cu/SAM/Au. Circular nanoparticles of Cu were deposited on SAM/Au while shattered like structure was obtained in case of Cu on gold substrate. The nucleation of Cu particles started in the defects and pinholes of SAM. Because the pinholes were wrapped by the hydrophobic alkyl chain of the hexanethiol SAM, the prolonged deposition time would increase the size of the Cu crystals. The deposited Cu growth would include the nucleation of isolated Cu and form circular microparticles [84]. Table 3.8 contains a brief summary of different SAMs modified electrodes utilized as glucose sensors.

Table 3.8. Different SAMs modified electrodes as glucose sensors.

Electrode	Linear range	Sensitivity	Detection limit	Ref.
Au/SAM of 4-mercaptobenzoic acid/glucose oxidase/Prussian Blue	12.5 µM-70 mM	50 nA/ cm^2 mM	12.5 µM	[83]
glucose oxidase/gold mercaptosuccinic anhydride self-assembled monolayer	0-120 mM	Not reported	Not reported	[85]
Cu particles/SAMs of 1-Hexanthiol modified gold electrode	3.0 µM to 10 mM	0.1116 A/M	0.7 µM	[84]
Glucose oxidase/self-assembled gold nanoparticles and double-layer 2D-network (3-mercaptopropyl)-trimethoxysilane polymer onto gold substrate	4.00×10^{-10} to 5.28×10^{-8} M	Not reported	1×10^{-10} M	[86]
Glucose oxidase/SAM of 3-mercaptopropioninc acid modified gold electrode	Not reported	Not reported	Not reported	[87]
Pt/glucose oxidase/SAM of 3-mercaptopropionic acid on gold	0-50 mM	1.5 µA cm^{-2} mM^{-1}	Not reported	[88]

3.5.9. Cholesterol Sensors

Cholesterol, a lipid found in all body cells, is needed to supply the fluidity and structural integrity of the cell membranes. It is considered as the precursor for the synthesis of different steroid hormones, sex hormones and vitamin D. Abnormal levels of cholesterol (>200 mg/dl) causes coronary heart disease, strokes and hardened arteries. Thus, it is very important to sensitively determine the concentration of cholesterol in serum and food for further diagnosis and treatment [40, 89]. SAMs can be utilized as a powerful substrate for the immobilization of various types of proteins and enzymes like cholesterol oxidase enzyme (ChOx) to be further utilized as cholesterol biosensor [40, 89-91]. In addition, the SAMs properties can be monitored through variation of the functional groups therefore the SAMs become appropriate for different biomolecules immobilization. Furthermore, SAMs may result in the orientation of the desired biomolecules without denaturation. Also, SAMs can facilitate the electron transfer between the studied biomolecules and the proposed surface [90].

B.D. Malhotra fabricated a cholesterol biosensor based on gold electrode modified with SAM of 11-amino-undecanethiol hydrochloride (AUT). The SAM modified Au was further modified through its terminal amino group with glutaraldehyde as a cross linking agent. Then cholesterol oxidase (ChOx) was covalently immobilized on the modified surface (Fig. 3.16) through the COOH terminal of glutaraldehyde. The proposed cholesterol biosensor was represented as ChOx/AUT/Au. The proposed sensor was utilized to detect cholesterol in the linear range of 50-500 mg/dL with sensitivity of 1.23 m^0/(mg dL) [40]. Table 3.9 contains a summary of different SAMs modified electrodes used as cholesterol biosensors.

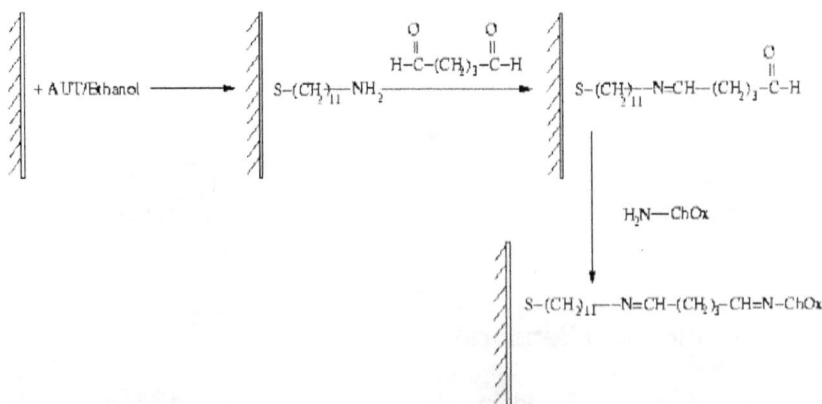

Fig. 3.16. Scheme of covalent immobilization of ChOx on AUT/Au SAM using glutaraldehyde as linking reagent, reused with permission [40].

Table 3.9. Different SAMs modified electrodes as cholesterol biosensors.

Electrode	Linear range	Sensitivity	Detection limit	Ref.
Cholesterol oxidase / SAM of Dimethyloctadecyl[3-(trimethoxysilyl)propyl]ammonium chloride (DMOAP) and (3-Aminopropyl)trimethoxy-silane (APTMS) based liquid crystal	10mg/dl to 250 mg/dl	Not reported	Not reported	[89]
Cholesterol oxidase / 11-amino-1-undecanethiol hydrochloride (AUT) self assembled monolayer on gold substrates	50 to 500 mg/dL	1.23 m^0/(mg dL	Not reported	[40]
Cholesterol oxidase /two-dimensional SAM of N-(2-aminoethyl)-3-aminopropyl-trimethoxysilane/indium–tin oxide coated glass plates	50 to 500 mg/dl	4.499×10^{-5} Abs (mg/dl)$^{-1}$	25 mg/dl	[90]
Au/SAM of cysteamine/gold nanoparticles/ Cholesterol oxidase	7.5×10^{-8} - 5×10^{-5} M	Not reported	5 ×10^{-9} M	[91]

References

[1]. L. M. Ghiringhelli, R. Caputo, L. D. Site, Alkanethiol headgroup on metal (111)-surfaces: general features of the adsorption onto group 10 and 11 transition metals, *Journal of Physics: Condensed Matter*, Vol. 19, 2007, pp. 176004-176014.

[2]. G. Liu, T. Böcking, J. J. Gooding, Diazonium salts: Stable monolayers on gold electrodes for sensing applications, *Journal of Electroanalytical Chemistry*, Vol. 600, 2007, pp. 335–344.

[3]. R. Dahint, Self-Assembled Monolayers–Principles and Applications, *SEACOAT*, 2010, Heidelberg.

[4]. I. Feliciano-Ramos, M. Caban-Acevedo, M. A. Scibioh, C. R. Cabrera, Self-assembled monolayers of L-cysteine on palladium electrodes, *Journal of Electroanalytical Chemistry*, Vol. 650, 2010, pp. 98–104.

[5]. C. R. Raj, S. Behera, Electrochemical studies of 6-mercaptonicotinic acid monolayer on Au electrode, *Journal of Electroanalytical Chemistry*, Vol. 581, 2005, pp. 61–69.

[6]. A. L. Eckermann, D. J. Feld, J. A. Shaw, T. J. Meade, Electrochemistry of redox-active self-assembled monolayers, *Coordination Chemistry Reviews*, Vol. 254, 2010, pp. 1769–1802.

[7]. P. Krysiński, M. Brzostowska-Smolska, Capacitance characteristics of self-assembled monolayers on gold electrode, *Bioelectrochemistry and Bioenergetics*, Vol. 44, 1998, pp. 163–168.

[8]. C. R. Raj, T. Ohsaka, Voltammetric detection of uric acid in the presence of ascorbic acid at a gold electrode modified with a self-assembled monolayer of heteroaromatic thiol, *Journal of Electroanalytical Chemistry*, Vol. 540, 2003, pp. 69–77.

[9]. C. Vericat, M. E. Vela, G. Benitez, P. Carro, R. C. Salvarezza, Self-assembled monolayers of thiols and dithiols on gold: new challenges for a well-known system, *Chemical Society Reviews*, Vol. 39, 2010, pp. 1805–1834.

[10]. D. Mandler, S. Kraus-Ophir, Self-assembled monolayers (SAMs) for electrochemical sensing, *Journal of Solid State Electrochemistry*, Vol. 15, 2011, pp. 1535–1558.

[11]. http://www.surface.mat.ethz.ch/research/surface_functionalization/alkane

[12]. M. Goldmann, J. V. Davidovits, P. Silberzan, Kinetics of self-assembled silane monolayers at various temperatures: evidence of 2D foam, *Thin Solid Films*, Vol. 327–329, 1998, pp. 166–171.

[13]. D. D. Erbahar, M. Harbeck, G. Gumus, I. Gurol, V. Ahsen, Self-assembly of phthalocyanines on quartz crystal microbalances for QCM liquid sensing applications, *Sensors and Actuators B*, Vol. 190, 2014, pp. 651–656.

[14]. C. M. Crudden, J. H. Horton, I. I. Ebralidze, O. V. Zenkina, A. B. McLean, B. Drevniok, Z. She, H. Kraatz, N. J. Mosey, T. Seki, E. C. Keske, J. D. Leake, A. Rousina-Webb, G. Wu, Ultra stable self-assembled monolayers of N-heterocyclic carbenes on gold, *Nature Chemistry*, 2014.

[15]. C. Nogues, P. Lang, Self-Assembled Alkanethiol Monolayers on a Zn Substrate: Structure and Organization, *Langmuir*, Vol. 23, 2007, pp. 8385-8391.

[16]. A. Galal, N. F. Atta, E. H. El-Ads, Probing cysteine self-assembled monolayers over gold nanoparticles – Towards selective electrochemical sensors, *Talanta*, Vol. 93, 2012, pp. 264–273.

[17]. E. H. El-Ads, Electrochemical sensor modified electrodes for the detection of some neurotransmitter compounds and pain reliever drugs, M. Sc. Thesis, *Cairo University,* 2012.

[18]. B. Arezki, A. Delcorte, P. Bertrand, Emission processes of molecule–metal cluster ions from self-assembled monolayers of octanethiols on gold and silver, *Applied Surface Science*, Vol. 231–232, 2004, pp. 122–126.

[19]. J. C. Love, D. B. Wolfe, R. Haasch, M. L. Chabinyc, K. E. Paul, G. M. Whitesides, R. G. Nuzzo, Formation and Structure of Self-Assembled Monolayers of Alkanethiolates on Palladium, *Journal of the American Chemical Society*, Vol. 125, 2003, pp. 2597-2609.

[20]. P. Lang, C. Nogues, Self-assembled alkanethiol monolayers on a Zn substrate: Interface studied by XPS, *Surface Science*, Vol. 602, 2008, pp. 2137–2147.

[21]. D. K. Aswal, S. Lenfant, D. Guerin, J. V. Yakhmi, D. Vuillaume, Self assembled monolayers on silicon for molecular electronics, *Analytica Chimica Acta*, Vol. 568, 2006, pp. 84–108.

[22]. A. Ulman, Formation and Structure of Self-Assembled Monolayers, *Chemical Reviews*, Vol. 96, Issue 4, 1996, pp. 1533–1554.

[23]. C. Vericat, M. E. Vela, G. A. Benitez, J. A. Martin Gago, X. Torrelles, R. C. Salvarezza, Surface characterization of sulfur and alkanethiol self-assembled monolayers on Au(111), *Journal of Physics: Condensed Matter*, Vol. 18, 2006, pp. R867–R900.

[24]. Y. Zhuo, R. Yu, R. Yuan, Y. Chai, C. Hong, Enhancement of carcinoembryonic antibody immobilization on gold electrode modified by gold nanoparticles and SiO_2/Thionine nanocomposite, *Journal of Electroanalytical Chemistry*, Vol. 628, 2009, pp. 90–96.

[25]. E. B. Bahadır, M. K. Sezgintürk, A comparative study of short chain and long chain mercapto acids used in biosensor fabrication: A VEGF-R1-based immunosensor as a model system, *Artificial Cells, Nanomedicine, and Biotechnology*, 2014.

[26]. A. Mirmohseni, J. Hosseini, M. Shojaei, S. Davaran, Design and evaluation of mixed self-assembled monolayers for a potential use in everolimus eluting coronary stents, *Colloids and Surfaces B: Biointerfaces*, Vol. 112, 2013, pp. 330–336.

[27]. D. W. M. Arrigan, L. Le Bihan, A study of l-cysteine adsorption on gold via electrochemical desorption and copper (II) ion complexation, *Analyst*, Vol. 124, 1999, pp. 1645–1649.

[28]. D. Merli, S. Protti, M. Labò, M. Pesavento, A. Profumo, A ω-mercaptoundecylphosphonic acid chemically modified gold electrode for uranium determination in waters in presence of organic matter, *Talanta*, Vol. 151, 2016, pp. 119–125.

[29]. N. F. Atta, A. Galal, E. H. El-Ads, A novel sensor of cysteine self-assembled monolayers over gold nanoparticles for the selective determination of epinephrine in presence of sodium dodecyl sulfate, *Analyst*, Vol. 137, 2012, pp. 2658–2668.

[30]. S. Campuzano, M. Pedrero, C. Montemayor, E. Fatàs, J. M. Pingarrón, Characterization of alkanethiol-self-assembled monolayers-modified gold electrodes by electrochemical impedance spectroscopy, *Journal of Electroanalytical Chemistry*, Vol. 586, 2006, pp. 112–121.

[31]. Y. Chen, L. Guo, W. Chen, X. Yang, B. Jin, L. Zheng, X. Xia, 3-mercaptopropylphosphonic acid modified gold electrode for electrochemical detection of dopamine, *Bioelectrochemistry*, Vol. 75, 2009, pp. 26–31.

[32]. A. Duwez, Exploiting electron spectroscopies to probe the structure and organization of self-assembled monolayers: a review, *Journal of Electron Spectroscopy and Related Phenomena*, Vol. 134, 2004, pp. 97–138.

[33]. T. Łuczak, Comparison of electrochemical oxidation of epinephrine in the presence of interfering ascorbic and uric acids on gold electrodes modified

with S-functionalized compounds and gold nanoparticles, *Electrochimica Acta*, Vol. 54, 2009, pp. 5863–5870.

[34]. S. Wang, D. Du, Q. Zou, Electrochemical behavior of epinephrine at L-cysteine self-assembled monolayers modified gold electrode, *Talanta*, Vol. 57, 2002, pp. 687–692.

[35]. X. Zhang, S. Wang, Voltametric Behavior of Noradrenaline at 2-Mercaptoethanol Self-Assembled Monolayer Modified Gold Electrode and its Analytical Application, *Sensors,* Vol. 3, 2003, pp. 61-68.

[36]. S. Aryal, B. K. C. Remant, N. Dharmaraj, N. Bhattarai, C. H. Kim, H. Y. Kim, Spectroscopic identification of S–Au interaction in cysteine capped gold nanoparticles, *Spectrochimica Acta Part A*, Vol. 63, 2006, pp. 160–163.

[37]. S. Wang, D. Du, Differential pulse voltammetry determination of ascorbic acid with ferrocene-l-cysteine self-assembled supramolecular film modified electrode, *Sensors and Actuators B*, Vol. 97, 2004, pp. 373–378.

[38]. Z. Nováková1, R. Oriňáková, A. Oriňák, P. Hvizdoš, A. S. Fedorková, Elimination Voltammetry as a New Method for Studying the SAM Formation, *International Journal of Electrochemical Science*, Vol. 9, 2014, pp. 3846 – 3863.

[39]. Q. Wang, D. Dong, N. Li, Electrochemical response of dopamine at a penicillamine self-assembled gold electrode, *Bioelectrochemistry*, Vol. 54, 2001, pp. 169–175.

[40]. P. R. Solanki, S. K. Arya, Y. Nishimura, M. Iwamoto, B. D. Malhotra, Cholesterol Biosensor Based on Amino-Undecanethiol Self-Assembled Monolayer Using Surface Plasmon Resonance Technique, *Langmuir,* Vol. 23, 2007, pp. 7398-7403.

[41]. L. S. Koodlur, Layer-by-layer self assembly of a water-soluble phthalocyanine on gold, Application to the electrochemical determination of hydrogen peroxide, *Bioelectrochemistry*, Vol. 91, 2013, pp. 21–27.

[42]. N. K. Chaki, K. Vijayamohanan, Self-assembled monolayers as a tunable platform for biosensor Applications, *Biosensors and Bioelectronics*, Vol. 17, 2002, pp. 1–12.

[43]. R. K. Shervedani, Z. Rezvaninia, H. Sabzyan, H. Z. Boeini, Characterization of gold-thiol-8-hydroxyquinoline self-assembled monolayers for selective recognition of aluminum ion using voltammetry and electrochemical impedance spectroscopy, *Analytica Chimica Acta*, Vol. 825, 2014, pp. 34–41.

[44]. R. S. Freire, L. T. Kubota, Application of self-assembled monolayer-based electrode for voltammetric determination of copper, *Electrochimica Acta*, Vol. 49, 2004, pp. 3795–3800.

[45]. W. P. See, L. Y. Heng, S. Nathan, Highly sensitive Aluminum (III) ion sensor based on a Self-assembled monolayer on a gold nanoparticles modified screen printed electrode, *Analytical Sciences*, Vol. 31, 2015, pp. 997-1003.

[46]. W. Yang, J. J. Gooding, D. B. Hibbert, Characterization of gold electrodes modified with self-assembled monolayers of L-cysteine for the adsorptive

stripping analysis of copper, *Journal of Electroanalytical Chemistry*, Vol. 516, 2001, pp. 10–16.

[47]. N. Yang, X. Wang, Q. Wan, Electrochemical reduction of Zn(II) ions on l-cysteine coated gold electrodes, *Electrochimica Acta*, Vol. 51, 2006, pp. 2050–2056.

[48]. N. F. Atta, S. M. Ali, E. H. El-Ads, A. Galal, Nano-perovskite carbon paste composite electrode for the simultaneous determination of dopamine, ascorbic acid and uric acid, *Electrochimica Acta*, Vol. 128, 2014, pp. 16–24.

[49]. M. Behpour, S. M. Ghoreishi, E. Honarmand, M. Salavati-Niasari, A novel N, N'-[1, 1'-Dithiobis(phenyl)] bis(salicylaldimine) self-assembled gold electrode for determination of dopamine in the presence of high concentration of ascorbic acid, *Journal of Electroanalytical Chemistry*, Vol. 653, 2011, pp. 75–80.

[50]. N. F. Atta, E. H. El-Ads, A. Galal, Self-Assembled Monolayers on Nanostructured Composites for Electrochemical Sensing Applications, in Handbook of Nanoelectrochemistry, *Springer*, 2015, pp. 417-478.

[51]. S. K. Moccelini, S. C. Fernandes, I. C. Vieira, Bean sprout peroxidase biosensor based on l-cysteine self-assembled monolayer for the determination of dopamine, *Sensors and Actuators B*, Vol. 133, 2008, pp. 364–369.

[52]. Y. Xian, H. Wang, Y. Zhou, D. Pan, F. Liu, L. Jin, Preparation of L-Cys–Au colloid self-assembled nanoarray electrode based on the microporous aluminium anodic oxide film and its application to the measurement of dopamine, *Electrochemistry Communications*, Vol. 6, 2004, pp. 1270–1275.

[53]. T. Tsai, F. Huang, J. J. Chen, Selective detection of dopamine in urine with electrodes modified by gold nanodendrite and anionic self-assembled monolayer, *Sensors and Actuators B*, Vol. 181, 2013, pp. 179-186.

[54]. R. K. Shervedani, M. Bagherzadeh, S. A. Mozaffari, Determination of dopamine in the presence of high concentration of ascorbic acid by using gold cysteamine self-assembled monolayers as a nanosensor, *Sensors and Actuators B*, Vol. 115, 2006, pp. 614–621.

[55]. T. Liu, M. Li, Q. Li, Electroanalysis of dopamine at a gold electrode modified with N-acetylcysteine self-assembled monolayer, *Talanta*, Vol. 63, 2004, pp. 1053–1059.

[56]. Q. Wang, N. Jiang, N. Li, Electrocatalytic response of dopamine at a thiolactic acid self-assembled gold electrode, *Microchemical Journal*, Vol. 68, 2001, pp. 77–85.

[57]. H. Zhang, N. Li, Z. Zhu, Electrocatalytic response of dopamine at a DL-homocysteine self-assembled gold electrode, *Microchemical Journal*, Vol. 64, 2000, pp. 277–282.

[58]. L. Codognoto, E. Winter, J. A. R. Paschoal, H. B. Suffredini, M. F. Cabral, S. A. S. Machado, S. Rath, Electrochemical behavior of dopamine at a 3,

3'-dithiodipropionic acid self-assembled monolayers, *Talanta*, Vol. 72, 2007, pp. 427–433.

[59]. C. R. Raj, T. Okajima, T. Ohsaka, Gold nanoparticle arrays for the voltammetric sensing of dopamine, *Journal of Electroanalytical Chemistry*, Vol. 543, 2003, pp. 127–133.

[60]. T. Tsai, C. Guo, H. Han, Y. Li, Y. Huang, C. Li, J. Jason Chen, Microelectrodes with gold nanoparticles and self-assembled monolayers for in vivo recording of striatal dopamine, *Analyst*, Vol. 137, 2012, pp. 2813–2820.

[61]. T. Łuczak, M. Bełtowska-Brzezinska, R. Holze, Electrocatalytic activity of gold modified with gold nanoparticles and self-assembled layers of meso-2, 3-dimercaptosuccinic acid for oxidation of epinephrine in the presence of ascorbic and uric acids, *Electrochimica Acta*, Vol. 123, 2014, pp. 135– 143.

[62]. L. Li, L. Yu, Q. Chen, H. Cheng, F. Wu, J. Wu, H. Kong, Determination of Ferulic Acid Based on *L*-Cysteine Self-assembled Modified Gold Electrode Coupling Irreversible Biamperometry, *Chinese Journal of Analytical Chemistry*, Vol. 35, Issue 7, 2007, pp. 933–937.

[63]. M. Mazloum-Ardakani, A. Dehghani-Firouzabadi, M. A. Sheikh-Mohseni, A. Benvidi, B. F. Mirjalili, R. Zare, A self-assembled monolayer on gold nanoparticles modified electrode for simultaneous determination of isoproterenol and uric acid, *Measurement*, Vol. 62, 2015, pp. 88–96.

[64]. G. Hu, Y. Ma, Y. Guo, S. Shao, Electrocatalytic oxidation and simultaneous determination of uric acid and ascorbic acid on the gold nanoparticles-modified glassy carbon electrode, *Electrochimica Acta*, Vol. 53, 200, pp. 6610–6615.

[65]. Y. Zhao, J. Bai, L. Wang, E. Xu Hong, P. Huang, H. Wang, L. Zhang, Simultaneous Electrochemical Determination of Uric Acid and Ascorbic Acid Using L-Cysteine Self-Assembled Gold Electrode, *International Journal of Electrochemical Science*, Vol. 1, 2006, pp. 363-371.

[66]. A. Sivanesan, P. Kannan, S. A. John, Electrocatalytic oxidation of ascorbic acid using a single layer of gold nanoparticles immobilized on 1, 6-hexanedithiol modified gold electrode, *Electrochimica Acta*, Vol. 52, 2007, pp. 8118–8124.

[67]. L. Zhang, R. Yuan, Y. Chai, X. Li, Investigation of the electrochemical and electrocatalytic behavior of positively charged gold nanoparticle and l-cysteine film on an Au electrode, *Analytica Chimica Acta*, Vol. 596, 2007, pp. 99–105.

[68]. V. Alvaro, R. Gutierrez, H. Martin, M. Bo, Screening of self-assembled monolayer for aflatoxin B1 detection using immune-capacitive sensor, *Biotechnology Reports*, Vol. 8, 2015, pp. 144–151.

[69]. H. C. W. Hays, P. A. Millner, M. I. Prodromidis, Development of capacitance based immunosensors on mixed self-assembled monolayers, *Sensors and Actuators B*, Vol. 114, 2006, pp. 1064–1070.

[70]. M. C. Canbaz, C. Şimşek, M. K. Sezgintürk, Electrochemical biosensor based on self-assembled monolayers modified with gold nanoparticles for detection of HER-3, *Analytica Chimica Acta*, Vol. 814, 2014, pp. 31–38.

[71]. H. Cho, J. Zook, T. Banner, S. Park, B. Min, K. A Hasty, E. Pinkhassik, E. Lindner, Immobilization of Fibrinogen Antibody on Self-Assembled Gold Monolayers for Immunosensor Applications, *Tissue Engineering and Regenerative Medicine*, Vol. 11, Issue 1, 2014, pp. 10-15.

[72]. X. Su, Y. Li, A self-assembled monolayer-based piezoelectric immunosensor for rapid detection of Escherichia coli O157:H7, *Biosensors and Bioelectronics*, Vol. 19, 2004, pp. 563–574.

[73]. M. K. Patel, P. R. Solanki, S. Khandelwal, V. V. Agrawal, S. G. Ansari, B. D. Malhotra, Self-assembled monolayer based electrochemical nucleic acid sensor for Vibrio cholerae detection, *Journal of Physics: Conference Series*, Vol. 358, 2012, pp. 012009-012017.

[74]. F. Kuralay, S. Campuzano, J. Wang, Greatly extended storage stability of electrochemical DNA biosensors using ternary thiolated self-assembled monolayers, *Talanta*, Vol. 99, 2012, pp. 155–160.

[75]. R. K. Shervedani, M. S. Foroushani, S. B. Dehaghi, Functionalization of gold mercaptopropionic acid self-assembled monolayer with 5-amino-1, 10-phenanthroline: Interaction with iron (II) and application for selective recognition of guanine, *Electrochimica Acta*, Vol. 164, 2015, pp. 344–352.

[76]. Meijia Wang, Chunyan Sun, Lianying Wang, Xiaohui Ji, Yubai Bai, Tiejin Li, Jinghong Li, Electrochemical detection of DNA immobilized on gold colloid particles modified self-assembled monolayer electrode with silver nanoparticle label, *Journal of Pharmaceutical and Biomedical Analysis*, Vol. 33, 2003, pp. 1117–1125.

[77]. T. H. Degefa, J. Kwak, Electrochemical impedance sensing of DNA at PNA self assembled monolayer, *Journal of Electroanalytical Chemistry*, Vol. 612, 2008, pp. 37–41.

[78]. A. J. Jeevagan, S. A. John, Electrochemical sensor for guanine using a self-assembled monolayer of 1, 8, 15, 22-tetraaminophthalocyanatonickel(II) on glassy carbon electrode, *Analytical Biochemistry*, Vol. 424, 2012, pp. 21–26.

[79]. G. Yang, R. Yuan, Y. Chai, A high-sensitive amperometric hydrogen peroxide biosensor based on the immobilization of hemoglobin on gold colloid/l-cysteine/gold colloid/nanoparticles Pt–chitosan composite film-modified platinum disk electrode, *Colloids and Surfaces B: Biointerfaces*, Vol. 61, 2008, pp. 93–100.

[80]. J. Jia, B. Wang, A. Wu, G. Cheng, Z. Li, S. Dong, A Method to Construct a Third-Generation Horseradish Peroxidase Biosensor: Self-Assembling Gold Nanoparticles to Three-Dimensional Sol-Gel Network, *Analytical Chemistry*, Vol. 74, Issue 9, 2002, pp. 2217-2223.

[81]. X. Li, J. Wu, N. Gao, G. Shen, R. Yu, Electrochemical performance of l-cysteine–goldparticle nanocomposite electrode interface as applied to

preparation of mediator-free enzymatic biosensors, *Sensors and Actuators B*, Vol. 117, 2006, pp. 35–42.

[82]. J. N. Patel, B. Kaminska, B. Gray, B. D. Gates, Effect of self-assembled monolayers (SAMs) in binding glucose oxidase for electro-enzymatic glucose sensor with gold electrodes, in *Proceedings of the 29th Annual International Conference of the IEEE EMBS (Cité Internationale)*, Lyon, France, 2007, pp. 2677-2680.

[83]. H. Wang, H. Ohnuki, H. Endo, M. Izumi, Effects of self-assembled monolayers on amperometric glucose biosensors based on an organic–inorganic hybrid system, *Sensors and Actuators B*, Vol. 168, 2012, pp. 249– 255.

[84]. J. Zhao, F. Wang, J. Yu, S. Hu, Electro-oxidation of glucose at self-assembled monolayers incorporated by copper particles, *Talanta*, Vol. 70, 2006, pp. 449–454.

[85]. R. K. Shervedani, A. Hatefi-Mehrjardi, Electrochemical characterization of directly immobilized glucose oxidase on gold mercaptosuccinic anhydride self-assembled monolayer, *Sensors and Actuators B*, Vol. 126, 2007, pp. 415–423.

[86]. X. Zhong, R. Yuan, Y. Chai, Y. Liu, J. Dai, D. Tang, Glucose biosensor based on self-assembled gold nanoparticles and double-layer 2d-network (3-mercaptopropyl)-trimethoxysilane polymer onto gold substrate, *Sensors and Actuators B*, Vol. 104, 2005, pp. 191–198.

[87]. J. J. Gooding, P. Erokhin, D. Losic, W. Yang, V. Policarpio, J. Liu, F. M. Ho, M. Situmorang, D. B. Hibbert, J. G. Shapter, Parameters important in fabricating enzyme electrodes using self-assembled monolayers of alkanethiols, *Analytical Sciences*, Vol. 17, 2001, pp. 3–9.

[88]. J. J. Gooding, V. G. Praig, E. A. H. Hall, Platinum-Catalyzed Enzyme Electrodes Immobilized on Gold Using Self-Assembled Layers, *Analytical Chemistry*, Vol. 70, 1998, pp. 2396-2402.

[89]. M. Tyagi, A. Chandran, T. Joshi, J. Prakash, V. V. Agrawal, A. M. Biradar, Self assembled monolayer based liquid crystal biosensor for free cholesterol Detection, *Applied Physics Letters*, Vol. 104, 2014, pp. 154104-154108.

[90]. S. K. Arya, A. K. Prusty, S. P. Singh, P. R. Solanki, M. K. Pandey, M. Datta, B. D. Malhotra, Cholesterol biosensor based on N-(2-aminoethyl)-3-aminopropyl-trimethoxysilane self-assembled monolayer, *Analytical Biochemistry*, Vol. 363, 2007, pp. 210–218.

[91]. N. Zhou, J. Wang, T. Chen, Z. Yu, G. Li, Enlargement of Gold Nanoparticles on the Surface of a Self-Assembled Monolayer Modified Electrode: A Mode in Biosensor Design, *Analytical Chemistry*, Vol. 78, 2006, pp. 5227-5230.

4.

Perovskites: Smart Nanomaterials for Sensory Applications

Nada F. Atta, Ekram H. El-Ads and Ahmed Galal

4.1. Introduction to Perovskite Oxides Nano-materials

Perovskite oxides, discovered by Gustav Rose in the Ural Mountains of Russia in 1839, have been named after the Russian mineralogist Perovski. The first discovered perovskite was the mineral $CaTiO_3$ [1-5]. There are several types of perovskites in our Earth's crust and some are more abundant like $FeSiO_3$ and $MgSiO_3$. The general chemical formula for the perovskites is ABC_3 with B ion surrounded by an octahedron of C ions. The most common formula for perovskites is ABO_3 with an alkali metal or a lanthanide with larger radii occupying the "A" site, a transition metal with small radii occupying the "B" site and oxygen ion exhibiting the ratio of 1:1:3 [1, 2, 5]. Hines et al described the perovskite structure as corner linked BO_6 octahedra with interstitial A cations [1-10]. The cubic form of ABO_3 perovskite contains "A" atom located at body center position, "B" atom located at cube corner position and oxygen atoms located at face centered positions (Fig. 4.1) [4]. Fig. 4.1 shows the structural representation of the ideal cubic perovskite showing the (a) cubic "A" unit cell and (b) cubic "B" unit cell. The cubic structure of perovskite is stabilized by the 6-fold coordination of "B" cation (octahedron) and the 12-fold coordination of the A cation [1-10]. Some distortions may occur in the ideal cubic structure of perovskite leading to different forms like orthorhombic, tetragonal, rhombohedral and hexagonal (Fig. 4.2) [3-7]. Fig. 4.3 shows the transformation of the perovskite from the ideal cubic structure to orthorhombic structure. Tilting the BO_6 octahedra and the displacement of the A cation results in maintaining the A and B and B site oxygen coordination in all perovskite distortions [4].

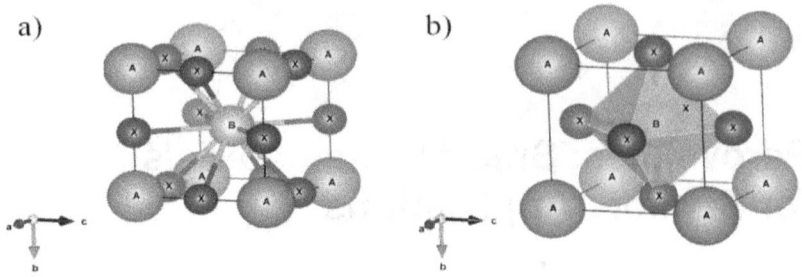

Fig. 4.1. Structural representation of the ideal cubic perovskite showing the (a) cubic A unit cell and (b) cubic B unit cell, reused with permission [7].

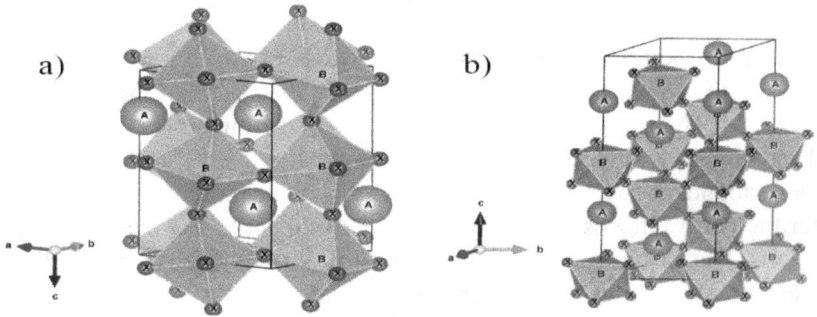

Fig. 4.2. Typical perovskite (a) orthorhombic and (b) hexagonal structural unit cells, reused with permission [7].

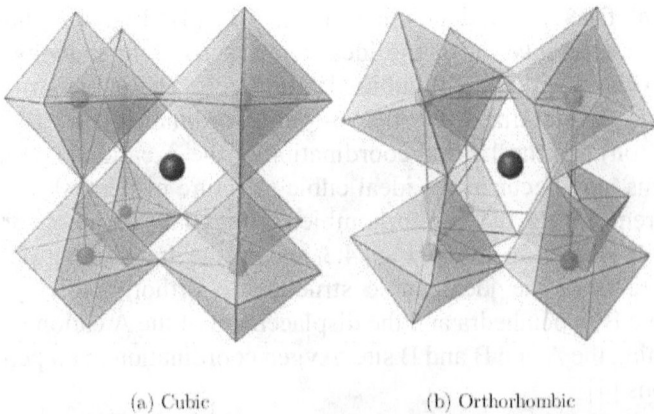

(a) Cubic   (b) Orthorhombic

Fig. 4.3. The transformation of the perovskite from the ideal cubic structure (a) to orthorhombic structure (b), reused with permission [4].

Two important rules must be verified for the perovskite formation; i) the charges of A and B ions should be equivalent to the whole charge of the oxygen ions therefore the whole charge is suitable for electrical neutrality of the perovskite formula ($A^{1+}B^{5+}O_3$, $A^{4+}B^{2+}O_3$ or $A^{3+}B^{3+}O_3$) and ii) the tolerance factor should be in the range of 0.8-1.0 with radii of A and B ions greater than 0.090 nm and 0.051 nm, respectively.

The tolerance factor "t", developed by V.M. Goldschmidt and described in equation 4.1, can be used to determine the convenience of the cations combination in the perovskite lattice [2-9].

$$t = (r_A + r_O) / [2^{1/2} (r_B + r_O)], \qquad (4.1)$$

where r_A, r_B and r_O are the radius of the A-site cation, the B-site cation and oxygen ion O^{2-}, respectively. The value of "t" approaches unity when the structure resembles the cubic form therefore it is considered as a real description of perovskite distortion from ideal cubic form [4]. When r_A decreases and/or r_B increases, the tolerance factor will decrease. Hines et al suggested that the structure of perovskite can be depicted by calculating the tolerance factor. The perovskite structure is hexagonal, cubic, orthorhombic and hexagonal ilmenite structure (like $FeTiO_3$) in the case of $1.00 < t < 1.13$, $0.9 < t < 1.0$, $0.75 < t < 0.9$ and $t < 0.75$, respectively [4].

Perovskite nano-oxides show very fascinating characteristics allowing them to be applied in various fields in large scale. They present electrical properties and diversity of solid state phenomena from insulating, semiconducting, metallic and superconducting properties [1, 2]. Perovskites nano-oxides present electrical, magnetic and optical properties due to the fact that more than 90 % of the periodic table elements can be stabilized in the perovskite structure and the chance of partial substitutions of different cations in the A and B sites resulting in the formation of $A_{1-x}A'_xB_{1-y}B'_yO_3$ [6].

Also, good properties of perovskites make them superior materials for several applications like photo-chromic, electro-chromic, image storage, switching, filtering and surface acoustic wave signal processing devices. In addition, they have been widely used as a catalyst for different reactions like hydrogen evolution reaction, nitrogen oxides and oxygen reduction reactions and carbon monoxide and hydrocarbons oxidation. Furthermore, they show great influence in several electrochemical applications such as sensing, biosensing, photo-electrolysis of water producing hydrogen and fuel cells [1, 2].

4.2. Models for Perovskites

4.2.1. The Ionic Models

The perovskites are highly ionic but they also exhibit some covalent character. The ionic model assumes that A and B cations lose their electrons and donate them in sufficient number to oxygen anions in order to obtain O^{2-} ions. The chemical valence for cation A is assumed to be +1 (K^{1+}), +2 (Ca^{2+}) or +3 (La^{3+}) while the ionic state of transition metal B is determined by charge neutrality. If we considered that the charges of A and B ions are q_A and q_B, respectively then $q_B = 6-q_A$ taking the three oxygen ions contributing with the factor of 6. The number of d electrons remaining can be determined from the electronic configuration once the charge state of the B ion is determined. In case of $SrTiO_3$, we have Sr^{2+} and O^{2-} therefore the titanium is Ti^{4+}. The electronic configuration of neutral Ti is [Ar] $3d^2 4s^2$. In case of Ti^{4+}, the outer four electrons are removed leaving the closed shell Ar core [Ar]. As we can observe that all the ions of $SrTiO_3$ have closed shell configuration (Sr^{2+} and O^{2-} have the [Kr] and [Ne] configurations, respectively). According to the ionic model, the material is considered an insulator when all of the ions have closed shell configuration. In case of WO_3, the W^{6+} has [Xe] core closed shell while for $NaWO_3$, the W^{5+} has a d^1 configuration. The perovskite may consider metallic conductor if the B ion retains d electrons and taking into consideration other factors. $NaWO_3$ is considered as good metal as the B ion exhibits the d^1 configuration. For $Na_x WO_3$, there will be (x)d electrons per unit cell where Na donates its electron and W ions donate the remaining electrons needed to form O^{2-} ions. We can imagine that we have $(1-x)W^{6+}$ and xW^{5+} ions distributed in a random or an ordered manner or each tungsten ion has an average valence of $(W^{(6-x)+})$. This picture can't be concluded from the ionic model and other considerations should be reported [1].

4.2.2. Madelung and Electrostatic Potentials

The ionic model could be applied to isolated or free ions but ions aren't isolated and interact in several ways. One of the most important interactions between ions is the electrostatic fields due to the charges on ions. The Madelung potential is the most important electrostatic effect between ions. The A and B cations are surrounded by the negatively charged oxygen ions. Therefore, the electrons orbiting the A and B ions experience repulsive electrostatic (Madelung) potentials. While the electrons orbiting the oxygen ions are surrounded by positively charged A and B cations experiencing attractive electrostatic (Madelung) potentials. The "site Madelung potentials" are defined as the electrostatic potentials at the different lattice sites due to all of the other ions. For perovskite, the Madelung potentials have very high values due to the large ionic charges. The Madelung potentials for the B sites are 30-50 eV. In case of $A^{2+}B^{4+}O_3^{2-}$, the site Madelung potentials are +45.6 eV, +19.9 eV and -23.8 eV for B, A and O, respectively.

The energies associated with the Madelung potentials resulted in the stability of the perovskite structure. The oxygen ions bind a pair of electrons due to the attractive potential at the oxygen sites. Two factors affecting the ability of oxygen ions to bind electrons, its site potential and electron affinity. The affinity of O^- for the second electron is actually positive which means that the second electron wouldn't be bound on a free oxygen ion. O^{2-} is stable in the perovskite lattice due to the attractive electrostatic (Madelung) potentials. On the other hand, a d electron is bound to a Ti^{4+} ion with ionization energy of -43 eV. Donation of an electron from Ti^{3+} to an O^- ion in $SrTiO_3$ would be energetically unfavorable in absence of repulsive site Madelung potential. The combination of site Madelung potential and ionization energy resulted in an effective binding energy of d electron with a value of $-43+45.6 = +2.6$ eV (unbound) for $SrTiO_3$ with the full ionic charges. Therefore, the Madelung potentials are responsible for the ionic configuration. An orbital centered on an ion has a finite radial extent therefore an electron in such an orbital would represent the electrostatic field over a distance close to the ionic radius. If we want to know the complete effect of the electrostatic field on the electron state, we should know the behavior of the field as a function of position near ion site.

The monopole term is just the site Madelung potential therefore the site Madelung potential results in a shift in the energy of an electron localized on the site. The higher-order multipoles (dipole, quadrupole, etc.) create an electrostatic field (with the point group symmetry of the site) which leads to a lifting of the orbital degeneracy.

The B ion site is affected by the cubic electrostatic field leading to the splitting of the fivefold degenerate d states into two groups: e_g and t_{2g}. The e_g group is doubly degenerate corresponding to the d orbitals having wave functions with angular symmetry of $(x^2-y^2)/r^2$ and $(3z^2-r^2)/r^2$. While the t_{2g} group is three fold degenerate corresponding to the d orbitals having wave functions with angular symmetry of (xy/r^2), (xz/r^2) and

(yz/r^2). On the other hand, the oxygen 2p states are affected by the axial electrostatic field and split into two states: $p\perp$ (where the 2p orbitals oriented perpendicular to B-O axis) and $p_|$ (where the 2p orbitals oriented parallel to B-O axis). The $p\perp$ is a doubly degenerate level while $p_|$ is a non-degenerate level. The lowest unoccupied state of the A cation is s state which is a spatially non-degenerate state with spherical symmetry at a site of cubic symmetry. Therefore, the energy of s state is affected by the monopole (Madelung potential) and unaffected by the other multipoles.

The particular ordering of levels can be explained according to the orientation of orbitals relative to the charge distributions on neighboring ions. The e_g and t_{2g} orbitals have lobes directed along and the B-O axis (into the negative charge clouds of oxygen ions directly) and perpendicular to the B-O axis between the negative oxygen ions, respectively. Therefore, the e_g states exert a greater repulsion and lie at a higher energy than the t_{2g} states. Similarly, the $p\perp$ states lie above the $p_|$ states when it is mentioned that the B ion cores appeared as positively charged centers.

In case of $SrTiO_3$, an insulating perovskite, the d states are completely empty while the p states are completely filled and the energy difference "E_g" between the t_{2g} and $p\perp$ states is equal to the energy gap. On the other hand, the metallic and semi-conducting materials have partially filled d orbitals. $NaWO_3$ or ReO_3 which are considered as good metallic conductors have a single electron in t_{2g} states (d^1 configuration).

In most cases, the primary valence and conduction bands of a perovskite are at energies lower than the energy bands involving the s states of A cation therefore the energy bands involving the s states of A cations are un-occupied. Therefore, the s states of A cation doesn't present an important role in shaping the electronic properties of perovskite. This conclusion does not overrule the role of A ion but the electrostatic potentials experienced by the A cation have a great impact on the energy of p-d valence and conduction bands. Also, the size of A cation is very crucial in determining the distortion of the crystal structure from the ideal cubic form. However, the orbitals of A cation may usually be omitted from the electronic structure calculations if we have a certain perovskite structure and the effective electrostatic potentials acting on B and O ions. This may lead to a major conceptual simplification when we consider that the electronic properties of the perovskites may arise from the BO_3 in the ABO_3. This simplification reveals that the electronic structure of

$BaTiO_3$ and $SrTiO_3$ could be essentially the same and the electronic structure of Na_xWO_3 could be independent of x. This doesn't reveal that the same properties are realized, but only the same available electronic states. As an example, the properties of WO_3, which is considered as an insulator, are completely different from those of $NaWO_3$ which is considered as good metallic conductor. However, as a first approximation the only effect of sodium ion is to donate the electrons occupying the t_{2g} states of tungsten ions [1].

4.2.3. Covalent Mixing

Besides the electrostatic interactions, the ions can interact because of the overlap of the electron wave functions. This will result in hybridization between the p and d orbitals and covalent bonds formation between the transition metal ions and the oxygen ions. The covalence in the perovskites is an important factor affecting nearly all of the physical and chemical properties of the perovskites. Therefore it isn't correct to say that the covalent mixing in insulating materials like $SrTiO_3$ is negligible. A cluster of atoms consisting of a transition metal ion and its octahedron of oxygen ions is considered to understand the covalent mixing. In case of ionic model, the wave functions are either pure d orbital or pure p orbital. In case of cluster, the wave functions are still d or p orbitals in character but there is a significant covalent mixing between the two orbitals.

The covalent mixing takes place due to the overlap between the d orbitals centered on the cation and the p orbitals on neighboring oxygen ions. There are two types of p-d overlap: sigma overlap (overlap between the e_g type d orbitals with $p_|$ type p orbitals) and pi overlap (overlap between the t_{2g} type d orbitals and $p_\perp$ orbitals). The overlap between t_{2g} and $p_|$ or between e_g and $p_\perp$ vanishes by symmetry [1].

4.3. General Characteristics of Perovskites

Perovskites family has superior characteristics such as piezoelectricity ($Pb(Zr, Ti)O_3$, $(Bi, Na)TiO_3$), ferroelectricity ($BaTiO_3$, $PdTiO_3$), catalytic activity ($LaCoO_3$, $LaMnO_3$, $BaCuO_3$) and magnetic properties ($LaMnO_3$, $LaFeO_3$, La_2NiMnO_6). Perovskites nano-oxides are known with their superconductivity ($Ba_2YCu_3O_7$, $La_{0.9}Sr_{0.1}CuO_3$, $HgBa_2Ca_2Cu_2O_8$), good ionic conductivity ($La(Ca)AlO_3$, $CaTiO_3$, $La(Sr)Ga(Mg)O_3$, $BaZrO_3$,

$SrZrO_3$, $BaCeO_3$), high electrical conductivity very close to metals (ReO_3, $SrFeO_3$, $LaCoO_3$, $LaNiO_3$, $LaCrO_3$) and mixed ionic and electronic conductivity [9].

Ferroelectric behavior is one of the most important properties of perovskites ($BaTiO_3$, $PdZrO_3$ and their doped compounds). There is a strong relation between the ferroelectric behavior of $BaTiO_3$ and its crystal structure. Upon increasing the temperature, $BaTiO_3$ was converted through three phase transitions from monoclinic to tetragonal then to cubic. The cubic structure of $BaTiO_3$ appeared well at higher temperature (> 303 K) and as a result it didn't show any ferroelectric behavior. $BaTiO_3$ exhibited large dipole moment generation in its crystal structure as it showed high dielectric constant depending on its crystal structure anisotropy [9].

In addition, superconductivity is an important property of perovskites. One of the high-temperature superconductors is Cu-based perovskites (like La-Ba-Cu-O). The presence of Cu in B site is responsible for the superconductivity. $YBa_2Cu_3O_7$, $Bi_2Sr_2Ca_2Cu_3O_{10}$ and $HgBa_2Ca_2Cu_3O_{8+\delta}$ are examples of high temperature superconductors with critical temperature of superconducting transition (Tc) of 130–155 K. Furthermore, $LaCoO_3$ and $LaMnO_3$ resemble perovskites with high electronic conductivity close to metals like Cu. As a result, they have been applied as cathodes in solid oxide fuel cells with superior hole conductivity of 100 S/cm. Doping the A site with other cations will result in increasing the electronic conductivity of the perovskites via enhancing the amount of mobile charge carriers resulting from charge reparations [9].

In addition, perovskites show high catalytic activity and chemical stability so that they have been widely studied in the catalysis of different reactions. They can be depicted as oxygen-activated catalyst and as models of active sites. The perovskites can be prepared from elements with unusual valence states or a high extent of oxygen deficiency and this is due to the high stability of the perovskite structure. Also, perovskites show large number of oxygen vacancies resulting in high surface activity to oxygen reduction ratio or oxygen activation and high catalytic activity. They can be applied as intelligent automobile catalyst, cleaning catalyst and automobile exhaust gas catalyst for various catalytic environmental reactions. Particularly, perovskites containing Cu, Co, Mn or Fe exhibited excellent catalytic activity toward the decomposition of NO at high temperature ($2NO \rightarrow N_2 + O_2$). The high catalytic activity is attributed to oxygen deficiency and the simple elimination of the surface oxygen in the form of a reaction product [9].

4.4. Doping of Perovskites

There are different factors affecting the characteristics and the catalytic activity of perovskites like synthesis method, calcination conditions (fuel, temperature, time, atmosphere, etc.) and partial substitution of A and/ or B sites. The partial or total substitutions on A and/or B sites affected greatly the catalytic activity of the perovskite due to the modification of the oxidation state, chemical state variation of the elements at A- and/or B-site, oxygen vacancies variation, oxygen lattice mobility and structural defects formation [11-13]. The basic characters of perovskites can be determined from the powerful bond between the B-site metal ions and the oxygen ions therefore, the B-site cation is accountable for the catalytic activity of the perovskites [14, 15]. The properties of the main metal B and the dopant one M will appear in the doped perovskite $AB_{1-y}M_yO_3$ upon the partial substitution of B-site cation with other metals M [15]. On the other hand, the unusual oxidation states of B-site cations can be stabilized by the presence of A site cation through the formation of crystal lattice vacancies under control resulting in different catalytic performances [14]. Different characteristics of ABO_3 perovskites can be changed upon doping the A and/or B sites like ionic and electronic conductivity, catalytic activity and chemical, physical and electronic characteristics. The doped perovskites can be utilized widely in various applications [16-19]. Many studies have been performed to use the doped perovskites in various applications because of the availability of hosting different cations with different sizes and charges in the A- and B- sites of these perovskites and the flexibility of composition. The stable perovskite lattice can accommodate multiple cationic substitutions under the control of two factors; i) keeping the Goldschmidt tolerance factor in the range of 0.75-1 and ii) preserving the electro-neutrality [12, 20, 21]. Thus, mutable amounts of various structural and electronic lattice defects can be created in the perovskite lattice due to their non-stoichiometry. As a result, the perovskite activity will be affected and the unusual valence states of different metal ions can be stabilized in the perovskite lattice [14, 20]. Also, the structural deformations from the ideal cubic structure of the perovskite will affect greatly some physical characteristics associated with structural characters of the perovskite-type oxides [22].

The catalytic activity of the perovskites can be determined from the type of the metal ion at the B sites and their partial substitutions. A wide spectrum for the variation of the catalytic, physical and chemical characteristics of the doped perovskites can be achieved by the B-site metal ion substitution with different metal ions M in the doped perovskite $AB_{1-y}M_yO_3$ [17, 20]. There is an obvious synergistic interaction between the perovskite crystal lattice and the metal ions dissolved in the crystal lattice upon doping. This will result in increased redox reaction and better catalytic activity of the prepared perovskite [23]. Upon doping the B-site, a dramatic change in the magnetic and transport properties of the ABO_3 perovskite has been achieved due to an ionic valence effect and/or an ionic size effect [17]. Furthermore, the perovskite stability and catalytic activity can be improved upon doping the B-site with transition metals especially noble metals in the ABO_3 perovskites [24]. In addition, the simple perovskite structure can be altered by the incubation of two various B ions with suitable different charge and size [5]. Different B-sites doped perovskites were mentioned in the literature showing enhanced catalytic properties like $LaNi_xCo_{1-x}O_3$ [12], $LaB_{0.9}Pd_{0.1}O_3$ [14], $LaMn_{1-x}PdxO_3$ [25], $BaFe_{1-x}Y_xO_{3-\delta}$ [16], $BaFe_{0.85}Cu_{0.15}O_{3-\delta}$ [21], $LaNi_{1-x}Fe_xO_3$ [13], $LaFe_{0.95-x}Co_xPd_{0.05}O_3$ [23], $LaCo_{0.95}Pd_{0.05}O_3$ [24] and $Sr_2Pd_{0.7}Au_{0.3}O_3$ [26].

Atta et al have studied the effect of doping the B-site "Pd^{2+}" of Sr_2PdO_3 with Au^{3+} and investigated its electronic properties and catalytic activity. Sr_2PdO_3 doped with Au, $Sr_2Pd_{0.7}Au_{0.3}O_3$, resulted in perfect electronic properties, stabilization of the cubic perovskite structure and enhanced catalytic activity compared to un-doped Sr_2PdO_3. This is attributed to the well dispersion of Au^{3+} ions in the perovskite framework [26].

On the other hand, the substitution of the A-site cation in the ABO_3 perovskites with various metal M ions, to produce $A_{1-x}M_xBO_3$, affected greatly the crystal structure and the physical, electronic, optical and catalytic properties of the prepared perovskite [27-32]. A-site doping resulted in the improvement of the flexible physical and chemical properties, catalytic activity and ionic and electronic conductivity of the prepared perovskite. Partial substitution of trivalent A-site cation like La^{3+} with bivalent cation of a lower oxidation state like Ca^{2+} or Sr^{2+}, which acted as electron acceptor on the A-site, resulted in the formation of oxygen vacancies and/or formation of a fraction of B cation in a higher valence state. In addition, higher free volume in the crystal lattice, enhanced content and mobility of surface adsorbed lattice oxygen and stabilized perovskite structure can be obtained. These achievements led to greater ionic conductivity and enhanced catalytic activity of the doped perovskites [18, 27, 29, 30, 32]. On the other hand, doping the A-site with cation of the same oxidation state but with smaller ionic radius led to distorted stabilized perovskite structure with three dimensional networks of the BO_6 octahedra and variation of the B–O–B bond angle and B–O bond length which further led to higher catalytic activity of the prepared perovskites [3, 18, 28]. Galal et al. proved that doping the A-site of $SrRuO_3$ with Ca^{2+} clearly enhanced the catalytic activity toward the hydrogen evolution reaction compared to the un-doped one. The presence of two different metal ions (Sr^{2+} and Ca^{2+}) in the A-site enhanced the interaction between the A and B metal ions (synergistic interaction) and improved the surface area and the structure stability of the prepared materials. In addition, the catalytic activity of the prepared perovskite increased by increasing the fraction of Ca-doped because of the increased surface area and the enhanced structure stability [3]. Fig. 4.4 showed an example of doped perovskite in which Sr^{2+} ions are substituted by Ca^{2+} ions in the crystal lattice of $SrTiO_3$ [33]. Several examples of A-site doped perovskites were mentioned in literature like $(La_{1-x}Sr_x)_yMnO_{3\pm\delta}$ [27], $La_{1-x}Ce_xGaO_3$, $La_{1-x}Pr_xGaO_3$ and $La_{1-x}Nd_xGaO_3$ [28], $La_{1-x}Ca_xMnO_3$ [29], $La_{1-x}Na_xMnO_{3+\delta}$, $La_{1-x}Ca_xMnO_{3+\delta}$ [30], $(Ba_{0.93}Fe_{0.07})TiO_3$ [31], $La_{1-x}Sr_xNiO_3$, $La_{1-x}Sr_xMnO_3$ [32] and $Sr_{1.7}Ca_{0.3}PdO_3$ [26, 34].

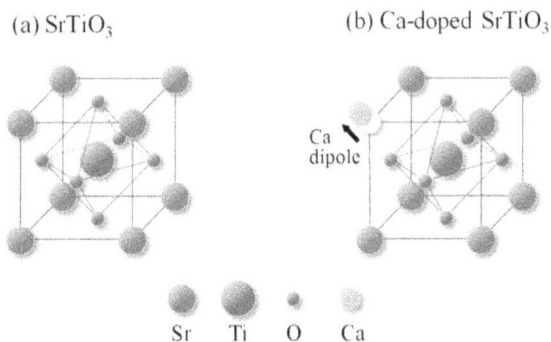

(a) $SrTiO_3$

(b) Ca-doped $SrTiO_3$

Ca dipole

Sr   Ti   O   Ca

Fig. 4.4. (a) Perovskite structure in $SrTiO_3$ and **(b)** Doped Ca ions are substituted for the Sr ions in Ca-doped $SrTiO_3$, reused with permission [33].

Atta et al studied the doping of A-site of Sr_2PdO_3 doped with Ca^{2+} ions and the electrocatalytic activity of $Sr_{2-x}Ca_xPdO_3$ (x = 0-0.7) toward the non-enzymatic glucose oxidation. Fig. 4.5 showed the effect of the dopant amount (Ca^{2+}) on the electrocatalytic activity of the proposed sensor graphite/$Sr_{2-x}Ca_xPdO_3$ (x = 0-0.7) toward non-enzymatic glucose sensing exhibiting a volcano shape with a maximum at x= 0.3. The anodic peak current of glucose increased reaching a maximum at 0.3 then decreased again at higher values of Ca. Doping the Sr_2PdO_3 with Ca^{2+} (x= 0.3) leads to an enhancement in the anodic peak current of glucose by 9 folds compared to the un-doped Sr_2PdO_3. This enhancement reflects the potential effect of Ca^{2+} as a dopant in the improvement of the electronic properties of the parent perovskite and enhancement of charge transfer rate. Higher free volume in the crystal lattice and enhanced content and mobility of surface adsorbed lattice oxygen are achieved upon doping the Sr^{2+} A-site with Ca^{2+} in Sr_2PdO_3 perovskite. In addition, a distorted stabilized perovskite structure is created upon doping leading to an obvious enhancement in the catalytic activity of the doped perovskites. Furthermore, the presence of two different metal ions in the A-site (Sr^{2+}, Ca^{2+}) would improve the synergistic interaction between the A and B metal ions and enhance the ionic and electronic conductivity, surface activity and the structure stability of the doped perovskites [26, 34].

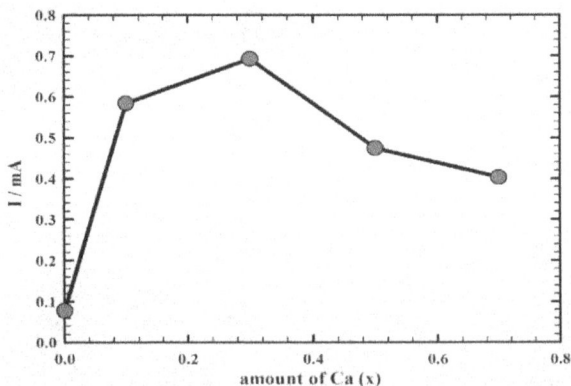

Fig. 4.5. The relationship between the dopant concentration (amount of Ca, x) and the anodic peak current of glucose in mA, reused with permission [26, 34].

4.5. Some Examples of Perovskites

4.5.1. Sr_2PdO_3

Strontium palladium or distrontium palladium perovskite, Sr_2PdO_3, is a superconducting material exhibiting a body-centered orthorhombic lattice structure as marked by ICDD [35]. Sr_2PdO_3 belongs to A_2BO_3 perovskite [36] and high T_c cuprate families and its prototype is Sr_2CuO_3 [37, 38]. The structure of Sr_2PdO_3 is shown in Fig. 4.6 (A) indicating the Sr^{2+} ions in the A site and Pd^{2+} ions in the B site. It is apparent that the crystal structure of Sr_2PdO_3 and Sr_2CuO_3 perovskites is the same [35, 39]. Sr_2CuO_3 with an orthorhombic unit cell resembles an intergrowth of an oxygen-deficient perovskite layer with a SrO rock salt layer [40]. In Sr_2PdO_3, Pd^{2+}, occupied the B-site and is surrounded by four coplanar (rectangular) oxygen atoms, while Sr^{2+} occupied the A-site and is surrounded by seven nearest oxygen neighbors [35, 36, 41]. Furthermore, Sr_2PdO_3 contains corner-linked chains of PdO_4 squares along the a-axis similar to the one-dimensional structure in Sr_2CuO_3 [36, 42].

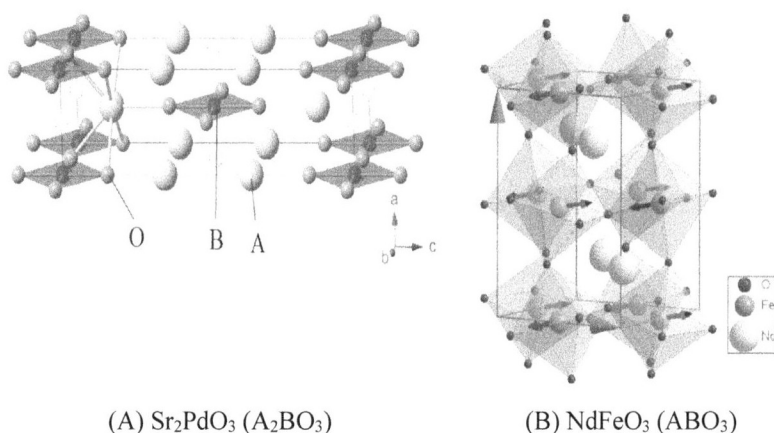

(A) Sr_2PdO_3 (A_2BO_3)   (B) $NdFeO_3$ (ABO_3)

Fig. 4.6. Structure of (A): Sr_2PdO_3, reused with permission [36] and (B): $NdFeO_3$, reused with permission [44].

Galal et al. prepared strontium palladium perovskite by citrate nitrate combustion method and proved its catalytic activity toward hydrogen evolution reaction with calculated activation energy of 27.9 kJ mol^{-1}. The presence of the modifier, strontium palladium, increased the current due to the hydrogen evolution 100 times compared to the unmodified electrode. In addition, the prepared strontium palladium perovskite exhibited an anti-ferromagnetic character with Neel temperature, $T_N = 370$ K [43].

4.5.2. NdFeO$_3$

NdFeO$_3$ "ABO$_3$", one of the rare-earth ortho-ferrites group, is an orthorhombic distorted perovskite where FeO$_6$ octahedra is formed by the coordination of Fe^{3+} ions by six O^{2-} [44-47]. The NdFeO$_3$ crystallographic unit cell is consisted of corner shared FeO$_6$ octahedra resulting in the formation of three-dimensional distorted perovskite structure (Fig. 4.6 (B)) [44]. Three competing magnetic interactions in NdFeO$_3$, Fe–Fe, Nd–Fe and Nd–Nd, determine its fascinating magnetic and structural properties resulting in various applications [45-47]. NdFeO$_3$ perovskites have been prepared by many methods like hydrothermal [48], combustion [49], sol-gel [50], precipitation [51] and sonication assisted precipitation [52].

NdFeO$_3$ was utilized as a superior catalyst for many reactions like hydrogen evolution reaction [53], methane combustion and CO oxidation [54]. Galal et al. investigated the electrocatalytic activity of NdFeO$_3$ which was prepared via microwave-assistant citrate method toward hydrogen evolution reaction. NdFeO$_3$ displayed superior electrocatalytic activity compared to GdFeO$_3$, LaFeO$_3$ and SmFeO$_3$ toward hydrogen evolution reaction in terms of low activation energy, high exchange current density, small strength of Fe–O bond which contains some degree of covalence and large inter-ionic distances (Fe–O) [53]. Also, it was utilized as an anode material for S/O$_2$ solid oxide fuel cells exhibiting good stability in sulfur vapor or SO$_2$ at 800 °C [55]. NdFeO$_3$ and its doped forms can provide many opportunities for improved gas sensing behavior [52, 56] like H$_2$S [45], CO gas [57], liquefied petroleum gas [52], hydrocarbon gases like methane, propane and n-hexane [56] and ethanol gas [45, 58]. NdFeO$_3$ nano-beads prepared by sonication assisted precipitation technique were utilized as a promising sensor for liquefied petroleum gas. The proposed sensor exhibited good reproducibility, high sensitivity, low response time, small recovery time, good sensing response, perfect stability and ease of preparation [52]. Sol-gel method was used to prepare NdFe$_{1-x}$Co$_x$O$_3$ (x= 0-0.5) perovskites to be examined as CO gas sensor. It was found that the largest cell volume was obtained in case of NdFeO$_3$. Also, there is a relation between the gas sensing properties of perovskites and the Fe-O bond strength. It was observed that the larger the bond length, the smaller the bond strength value and the higher the sensing response. NdFeO$_3$ has the largest bond length, the smallest bond strength and the highest sensing response toward CO gas compared to other NdFe$_{1-x}$Co$_x$O$_3$ [57].

4.6. Methods of Perovskite Synthesis

4.6.1. Citrate–nitrate Combustion Methods

Combustion method is considered as a suitable easy method for the preparation of various ceramics, nanomaterials and catalysts. A thermally induced redox reaction takes place between the oxidant and the fuel in this method. The powder obtained from this method is nano-sized, highly reactive and homogenous. When compared with other conventional methods such as solid state synthesis, a single phase perovskite is produced at lower temperatures or shorter reaction times. Citrate-nitrate combustion synthesis method is a very famous method in which the citric acid is considered as fuel and metal nitrates are considered as a source for metal and oxidant. Citrate-nitrate combustion synthesis method is similar to the well-known Pechini process "sol–gel combustion method" except that polyhydroxy alcohols like ethylene glycol are not utilized. Also, the citrate method is different from the Pechini process in that the nitrates aren't removed as NO_x but remain with the metal citrates in the mixture causing auto-combustion [59-69]. Iron, cobalt and cerium-perovskite can be prepared using citrate-nitrate combustion method [59]. Also, $La_{0.6}Sr_{0.4}CoO_{3-\delta}$ nano-powders can be prepared via combined citrate–EDTA method using citric acid, EDTA and metal nitrates as the precursor solution and calcination at 900 °C [60]. In addition, a dense $La_{0.8}Sr_{0.2}Co_{0.2}Fe_{0.8}O_{3-\delta}$ perovskite was prepared by citrate method using citrates as organic salts and nitrates as cation source and calcination at 900 °C [61]. Furthermore, it is possible to prepare Sr- or Ce-doped $La_{1-x}M_xCrO_3$ (x= 0.0, 0.1, 0.2 and 0.3) catalysts were prepared by thermal decomposition of citrate precursors followed by calcination at 800 °C in air [62].

On the other hand, the Pechini or citrate gel method involves two chemical reactions: i) chelation or complexation between metal ions and citric acid and ii) polyesterification of complexes with ethylene glycol. The previous two steps preserve the homogeneity of the metal salt in a gel. Also, this results in the formation of three-dimensional structures and minimizes the segregation. This method offers the advantage of high purity and good control of the composition of the resulting materials. Citric acid gel process was used to prepare lanthanum manganite oxides and as the calcination temperature increases from 600 °C to 1000 °C, the specific surface area decreases from 13 to 1.5 m^2 g^{-1} [63, 64]. Also, nanophasic $LaCoO_3$ thin films were prepared from precursor mixed solutions of citric acid and ethylene glycol with La and Co ions via Pechini-type reaction route through calcination at 600°C for 3 h [65-67]. Also, pure and Sr-doped LaGaO$_3$, LaFeO$_3$ and LaCoO$_3$ and Sr, Mg-doped LaGaO$_3$ were prepared using citrate-gel methods [68]. In addition, highly crystalline and homogeneous oxides of LaNiO$_3$ were prepared using sol–gel methodology (citrates method) [69].

4.6.2. Co-precipitation Method

Co-precipitation method involves oxalates precipitation from metal nitrates solution at pH< 1 followed by thermal decomposition forming oxide compounds. Another alternative of co-precipitation method is the co-precipitation of metal nitrates as hydroxides using a precipitating agent of ammonia in an alkaline pH range. Several factors affecting the properties of the prepared single phase should be controlled like pH, mixing rates, temperature and concentration. Highly crystalline BaZrO$_3$ powders can be prepared using co-precipitation in highly basic aqueous solution [70]. On the other hand, LaCoO$_3$ was prepared via the simultaneous oxidation/co-precipitation of an equimolar mixture of La(III) and Co(II) nitrates leading to the formation of hydroxide-containing gel followed by calcination at 600 °C [71]. Also, carbonate co-precipitation method was used to prepare La$_{0.8}$Sr$_{0.2}$Ga$_{0.8}$Mg$_{0.2}$O$_{2.8}$ powders [72]. Furthermore, co-precipitation method using ammonium carbonate or ammonium bicarbonate precipitant was used to prepare Sr- and Mg-doped LaGaO$_3$ (LSGM) and Co-doped LSGM (LSGMC) powders [73].

4.6.3. Microwave Synthesis

Microwave irradiation process (MIP), which was evolved from microwave sintering, has been applied in food drying, inorganic/organic synthesis, microwave-induced catalysis and plasma chemistry. MIP shows many advantages like (i) rapid reaction velocity; (ii) uniform heating; (iii) clean and energy efficient. The conditions used for microwave preparation are 2.45 GHz and with a maximum output power not less than 1 kW. Dielectric materials can absorb the microwave energy and transform it into heat energy directly through the polarization and dielectric loss in the interior of materials. 80–90 % is the energy utilization efficiency that can be reached through MIP and much higher than that reported via conventional routes [74]. A large number of perovskites like GaAlO$_3$, LaCrO$_3$ with fascinating characteristics such as ferroelectricity, superconductivity, high-temperature ionic conductivity and magnetic ordering variety have been prepared using microwave assisted method [75-77]. Perovskites with smaller grain size and more rapid lattice diffusion were prepared using microwave leading to enhanced lattice oxygen mobility in the catalytic processes [78]. Calcium titanate ($CaTiO_3$) powders can be prepared by microwave assisted method showing rapid structural organization than that prepared using other methods [79]. La–Ce–Mn–O catalysts were prepared using two different hydrothermal methods; conventional and dielectric heating. Microwave irradiation results in the formation of $La_{1-x}Ce_xMnO_{3+\varepsilon}CeO_2$ ($x+\varepsilon=0.2$) while the conventional heating results in the formation of $LaMnO_3+CeO_2$. $La_{1-x}Ce_xMnO_{3+\varepsilon}CeO_2$ prepared by microwave exhibited high sulfur tolerance and the best percent CH_4 conversion. This is attributed to the well dispersion of cerium oxide and its strong interaction with the $LaMnO_3$ structure [80].

Nano-sized pure single phases of perovskite type oxides like $LaFeO_3$, $SmFeO_3$, $NdFeO_3$ and $GdFeO_3$ can be prepared using microwave irradiation (2.45 GHz, 900W) resulting in finer particles and higher surface area compared to that prepared using other methods [81]. Barium iron niobate pure perovskite phase with a cubic symmetry was successfully synthesized using microwave method followed by calcination at 850 °C for 14 h [82]. In addition, one-dimensional single crystalline $KNbO_3$ nanostructures were prepared through the reaction of Nb_2O_5 and KOH under microwave-assisted hydrothermal synthesis [83]. Also, nanocrystalline $PbWO_4$ [84], $CaMoO_4$ [85] and MWO_4 (M: Ca, Ni) [86] materials were synthesized using modified citrate complex method assisted by microwave irradiation at low temperatures. As well, microwave-induced combustion process was used to prepare strontium hexaferrite nano-particles [87].

4.7. Characterization of Perovskites

The different phases of the prepared perovskites can be differentiated using X-ray powder diffraction, XRD. In addition, the structure of the perovskite can be characterized using single-crystal XRD analysis. Thermal analysis techniques like TGA, DTA and DSC can be used to test the thermal stability of the prepared perovskites. Furthermore, the different morphological and surface characteristics of the prepared perovskites can be identified using scanning (SEM) and transmission (TEM) electron microscopies. Also, surface area measurement of the prepared perovskites can be studied using BET. In addition, the formed phases of the prepared perovskites can be identified using FTIR "Fourier Transform infrared spectroscopy" and XPS "X-ray photo-electron spectroscopy" [10, 88-90].

4.7.1. XRD

XRD can be utilized for determination of the relative percent of different phases and identification of different phases of the prepared materials. XRD data can be used to calculate the different structural parameters like lattice parameters (a, b and c), lattice volume, crystal size and theoretical density. In addition, the optimum preparation conditions of different perovskites can be chosen based on XRD analysis [3, 89-91].

In XRD, diffraction occurs only when Bragg's Law is satisfied (equation (4.2)):

$$2d \sin\theta = n \lambda, \tag{4.2}$$

where d is the spacing between diffracting planes or atomic layers, θ is the incident angle, n is an integer number and λ is the wavelength of the X-ray beam which equals to 1.54 $°A$. The crystallite size can be calculated using Scherrer equation (equation (4.3)):

$$D = K \lambda / \beta \cos\theta, \tag{4.3}$$

where D is the mean crystallite size, K is the dimensionless shape factor which equals to 0.94 and changes as the actual shape of the crystallite changes, β is the full width at half the maximum intensity (FWHM) in radians and θ is the Bragg angle. Scherrer equation is applied only for nano-scale crystallites not for grains larger than 0.1 μm. The lattice parameters, a, b and c are related to the value of the spacing between diffracting planes (d) and miller indices (h, k, l) in case of orthorhombic crystal structure through the following equation (equation (4.4)):

$$1/d^2 = h^2/a^2 + k^2/b^2 + l^2/c^2 \tag{4.4}$$

On the other hand, the lattice volume (V / $°A^3$) can be calculated from the following equation (equation (4.5)):

$$V = abc \tag{4.5}$$

Moreover, the theoretical density (D_{theo} / g cm^{-1}) can be calculated from the following equation (equation (4.6)):

$$D_{theo} = ZM / N_A V, \qquad (4.6)$$

where Z is a parameter related to the crystal structure, M is the molecular weight of the prepared perovskite, N_A is Avogadro's number and V is the lattice volume [3].

Atta et al prepared $NdFeO_3$ perovskite via microwave assistant-citrate method and sintering for 5 hours at 900 °C [53]. In the prepared perovskite, the A-site was occupied by Nd^{3+} and B-site was occupied by Fe^{3+}. The phase formation of the prepared material was investigated using XRD. The tolerance factor of the prepared perovskite was 0.9232 which was close to unity reflecting the stability of the prepared perovskite. The ionic radii of Fe^{3+} (high spin), Nd^{3+} and O^{2-} were 0.645 A°, 1.27 A° and 1.40 A°, respectively. The XRD of $NdFeO_3$ perovskite was shown in Fig. 4.7 and the data was compared with the ICDD card of $NdFeO_3$ (card number 04-006-8303).

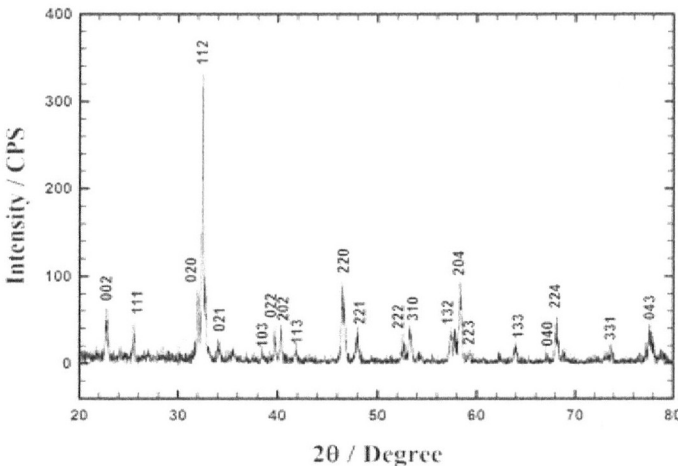

Fig. 4.7. XRD patterns of $NdFeO_3$ prepared by microwave assistant-citrate method. Miller indices (h, l, k) are written in black line, reused with permission [26, 92].

The XRD showed pure single phase of $NdFeO_3$ without any secondary phases with the major diffraction peak (112) which was matched well

167

with the theoretical one. The structure of NdFeO₃ perovskite was orthorhombic and the peak intensity was high suggesting high crystallinity of the prepared material. The crystal size of the prepared perovskite was 36.80 nm calculated using Scherrer equation. In addition, some structural parameters like theoretical density, lattice volume and lattice parameters were calculated from XRD data and summarized in Table 4.1. They were in a good matching with the theoretical values of the standard NdFeO₃ perovskite [26, 92].

Table 4.1. Structural parameters calculated from XRD data for NdFeO₃, reused with permission [26, 92].

	Crystal Structure	Lattice Parameters $A°$	Lattice Volume $A^{°3}$	Theoretical density g/cm^3	Crystal size nm	Goldsch midt tolerance factor
Standard NdFeO₃ (card number 04-006-8303)	Orthorh ombic	a = 5.4530 b = 5.5840 c = 7.7680	236.53	6.97	---	---
NdFeO₃ Citrate-nitrate method	Orthorh ombic	a = 5.4326 b = 5.5834 c = 7.7885	236.25	6.97	36.80	0.9232

4.7.2. Scanning Electron Microscopy and Transmission Electron Microscopy (SEM, TEM)

The morphology and surface characteristics of the perovskite nanomaterials can be investigated using SEM and TEM. The SEM of the prepared perovskites can be greatly affected by the preparation conditions, synthesis method, type of A and B-site metal ions and doping A- and/or B-sites [53, 90-96].

The catalytic activity of the prepared materials is highly affected by porosity, particle size and edges of perovskite crystals. Atta et al prepared Sr_2PdO_3 and $Sr_{1.7}Ca_{0.3}PdO_3$ by glycine-nitrate combustion method and examined the difference in their morphology via SEM. Fig. 4.8 (A and B) showed the SEM of Sr_2PdO_3 and $Sr_{1.7}Ca_{0.3}PdO_3$, respectively. Clusters of particles rather than separated ones are shown in the SEM of both perovskite materials. This is attributed to the agglomeration of the crystals over the surface of the graphite substrate. Both perovskites are homogenous in shape and size and well-embedded over the graphite surface [26, 34].

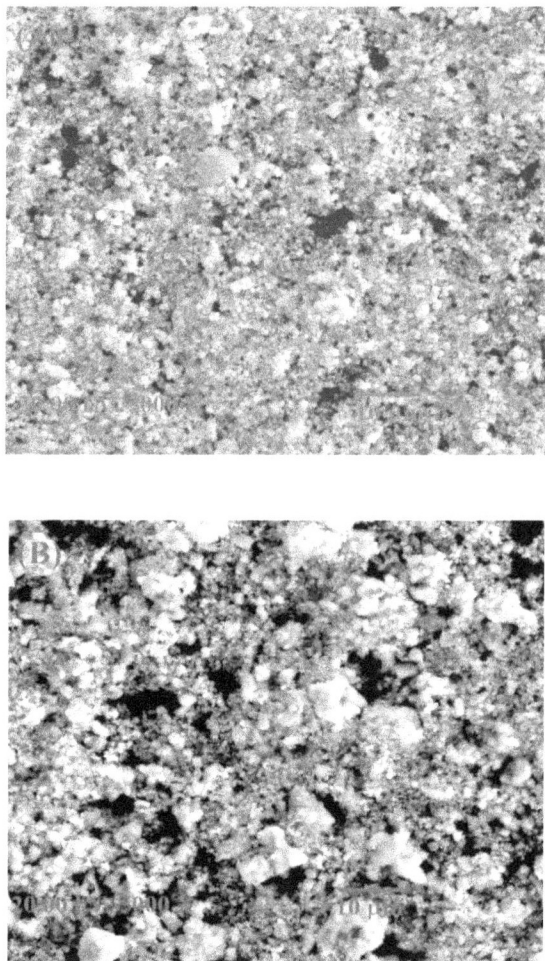

Fig. 4.8. SEM of Sr_2PdO_3 and $Sr_{1.7}Ca_{0.3}PdO_3$, reused with permission [34].

Atta et al prepared Sr_2PdO_3 and $Sr_{1.7}Ca_{0.3}PdO_3$ and confirmed the Ca insertion in the perovskite lattice by EDAX measurements. The EDAX measurements of Sr_2PdO_3 and $Sr_{1.7}Ca_{0.3}PdO_3$ are shown in Fig. 4.9 (A, B), respectively. The EDAX of Sr_2PdO_3 showed only Sr, Pd and O while EDAX of $Sr_{1.7}Ca_{0.3}PdO_3$ showed Sr, Ca, Pd and O confirming the

Ca doping in the Sr_2PdO_3 perovskite lattice. In addition, EDAX measurement can be used to confirm the atomic percent of Ca dopant. The atomic percent obtained from EDAX is 4.4 % which was close to its value in $Sr_{1.7}Ca_{0.3}PdO_3$ (5 %). The atomic percents of Sr, Ca, Pd and O in $Sr_{1.7}Ca_{0.3}PdO_3$ were 28.3 %, 5 %, 16.6 % and 50 %, respectively. The atomic percent of Ca obtained from EDAX was comparable with its real value in the Ca preparation precursor we started with (6 %). We can say that Ca in case of $Sr_{1.7}Ca_{0.3}PdO_3$ was completely inserted in the perovskite lattice and Ca^{2+} had substituted for Sr^{2+} in this sample [26, 34].

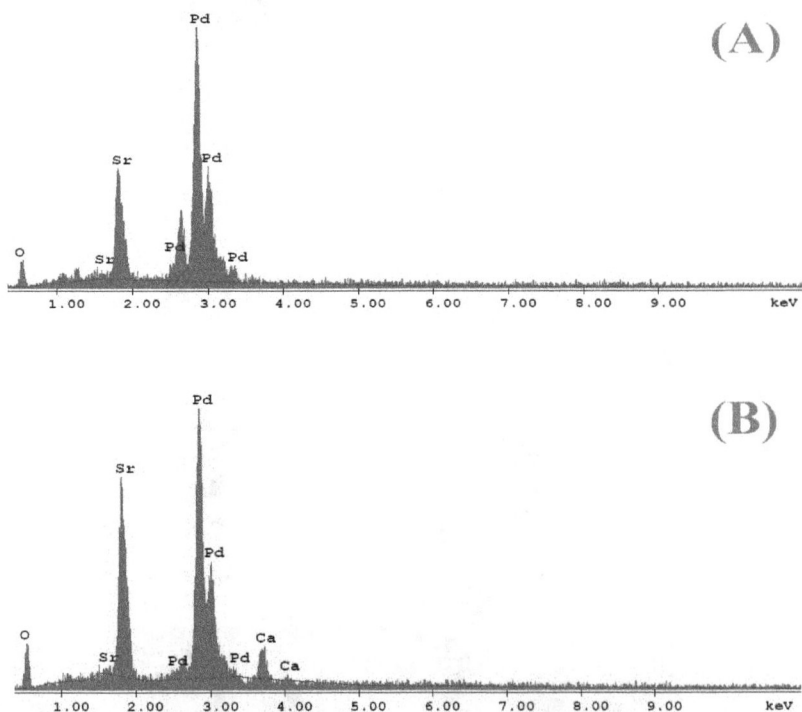

Fig. 4.9. EDAX analysis for Sr_2PdO_3 (A) and $Sr_1.7Ca_{0.3}PdO_3$ (B), reused with permission [26, 34].

Fig. 4.10 (A and C) showed the TEM images of Sr_2PdO_3 and $Sr_{1.7}Ca_{0.3}PdO_3$, respectively. The TEM diffraction patterns of Sr_2PdO_3 and $Sr_{1.7}Ca_{0.3}PdO_3$ are shown in Fig. 4.10 B and D, respectively. TEM textures in terms of distortion in shape, edges, roughness and defects affected greatly the catalytic activity of the prepared perovskites. Higher catalytic activity toward glucose sensing was achieved at $Sr_{1.7}Ca_{0.3}PdO_3$ compared to Sr_2PdO_3 due to the smaller crystals size, more edges and higher consistency in the shape of the $Sr_{1.7}Ca_{0.3}PdO_3$ crystals. The size of the crystals was in the range of 8-16 nm in case of $Sr_{1.7}Ca_{0.3}PdO_3$ compared to 15-29 nm in case of Sr_2PdO_3 leading to higher surface activity [26, 34].

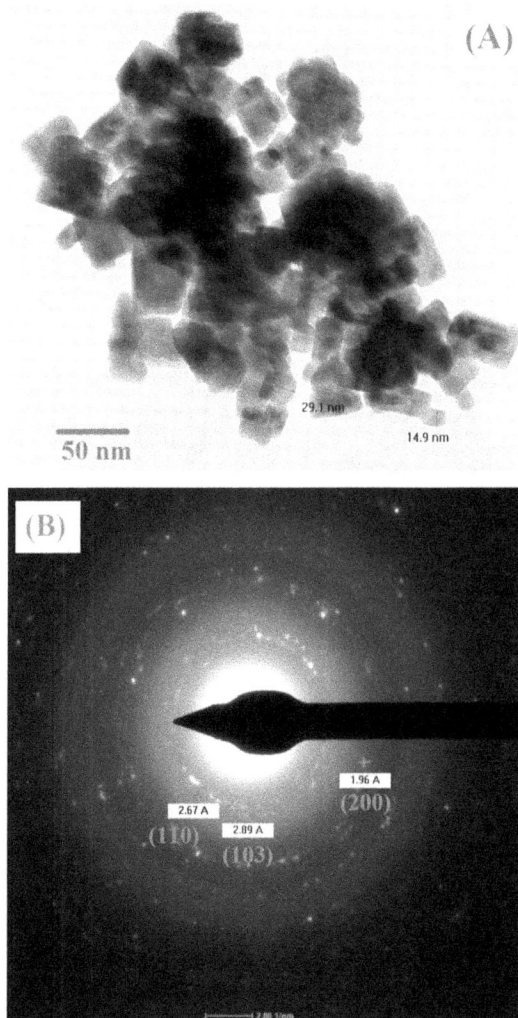

Fig. 4.10 (A, B). TEM micrographs of (A) Sr_2PdO_3 and (B) its diffraction pattern, reused with permission [26, 34].

Fig. 4.10 (C, D). TEM micrographs of (C) $Sr_{1.7}Ca_{0.3}PdO_3$ and (D) its diffraction pattern, reused with permission [26, 34].

4.7.3. Brunauer Emmett Teller Measurements (BET)

It is very important to measure the specific surface area of the prepared materials as the electrocatalytic activity and the electrochemical performance of the perovskites are greatly related to the specific surface area of the materials. Brunauer–Emmett–Teller (BET) nitrogen adsorption can be used to measure the surface area values of different perovskites. The surface area of the prepared perovskites is greatly affected by the preparation conditions, synthesis method, type of A and B-site metals and presence of different dopants [88, 89, 97].

Galal et al prepared lanthanide-transition metal perovskites, $LaBO_3$, with different types of the transition metal "B" like Ni, Fe, Mn, Cr and Co via microwave assistant-citrate method. Changing the type of transition metal in $LaBO_3$ perovskite affects greatly the surface area and the catalytic activity of the prepared perovskite toward oxygen evolution reaction. The values of the surface area measured by BET for the different prepared perovskites are summarized in Table 4.2. The order of decreasing the catalytic activity toward oxygen evolution reaction is as following; $LaNiO_3$ > $LaFeO_3$ > $LaMnO_3$ which matches well with the order of decreasing the surface area (25.0, 6.7 and 2.5 m^2/g, respectively). $LaNiO_3$ perovskite shows the highest surface area with the smallest particle size and thus the highest catalytic activity toward oxygen evolution reaction [88].

Table 4.2. Comparison between the particle size calculated from XRD and that measured by TEM in addition to the surface area measured value for each prepared perovskite, reused with permission [88].

Sample	Lattice structure from XRD	Particle size (L) calculated from XRD (nm)	Particle size (L) measured by TEM (nm)	Surface area by BET m^2/g
$LaNiO_3$	rhombohedral	21.4	26.3	25
$LaFeO_3$	orthorhombic	80.6	81.9	6.7
$LaMnO_3$	rhombohedral	45.2	57.9	2.5
$LaCrO_3$	orthorhombic	52.3	63.2	10.3
$LaCoO_3$	rhombohedral	136.2	240.0	1.4

4.7.4. Thermal Analysis

The thermal stability and the decomposition temperature of the prepared perovskites can be identified using thermal analysis. Thermal analysis can be used to identify the optimum calcination temperature of any perovskite [90, 98, 99].

Atta et al prepared Sr_2PdO_3 and Ca-doped Sr_2PdO_3 "$Sr_{1.7}Ca_{0.3}PdO_3$" perovskites by glycine-nitrate combustion method. The optimum calcination temperature of the precursor powders of $Sr_{1.7}Ca_{0.3}PdO_3$ was obtained from DTG curves. The DTG spectrum of Sr, Pd and Ca mixed glycine complex was performed and compared with the DTG of Sr and Pd mixed glycine complex (Fig. 4.11). A smooth weight loss step due to the decomposition of glycine complex was obtained at ~540 °C in case of $Sr_{1.7}Ca_{0.3}PdO_3$ and at ~350 °C in case of Sr_2PdO_3. A sharp weight loss was obtained in both cases at ~750 °C confirming the formation of $Sr_{1.7}Ca_{0.3}PdO_3$ and Sr_2PdO_3, respectively. As a result, glycine-nitrate combustion method was utilized for the synthesis of Sr_2PdO_3 and $Sr_{1.7}Ca_{0.3}PdO_3$ by calcination at 750 °C for 3 hours [26, 34].

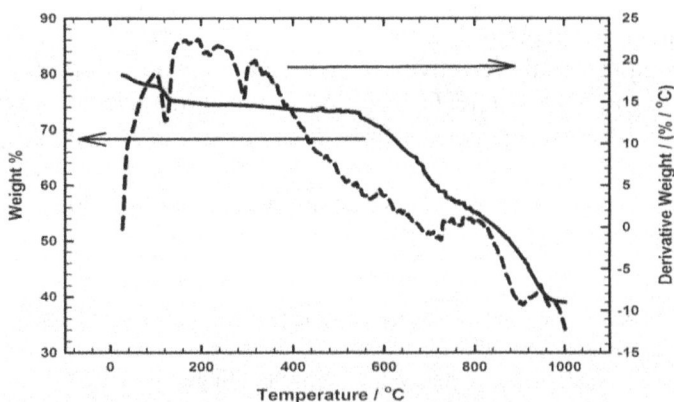

Fig. 4.11. The DTG spectrum of Sr, Pd and Ca mixed glycine complex in the temperature range from room temperature to 1000 °C in air with a heating rate of 5 °C/min, reused with permission [26, 34].

4.7.5. FTIR

FTIR can be used to examine the chemical structure and chemical bonding of the prepared perovskites. Structural confirmation similar to that obtained via XRD can be achieved using FTIR [89, 99-102]. Atta et al prepared $NdFeO_3$ perovskite by microwave assistant-citrate method and its structure was confirmed using FTIR. Fig. 4.12 showed the FTIR of NdFeO$_3$ perovskite in the frequency range from 4000 to 400 cm^{-1}. The two strong absorptive bands at about 550 cm^{-1} and 410 cm^{-1} were corresponding to the stretching vibration band of Fe–O and the bending vibration band of O–Fe–O in NdFeO$_3$ perovskite, respectively. These bands confirmed the formation of NdFeO$_3$ perovskite and this was comparable with XRD data [26, 92, 103, 104].

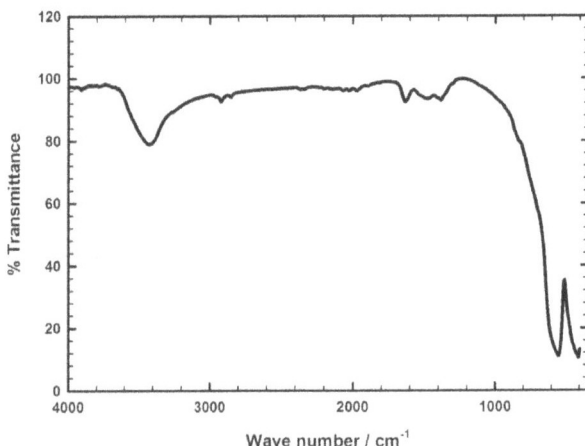

Fig. 4.12. FTIR spectra of NdFeO$_3$ perovskite in the frequency range from 4000 to 400 cm^{-1}, reused with permission [26, 92].

4.7.6. X-ray Photoelectron Spectroscopy (XPS)

XPS can be used to identify the surface compositions of the various components of the prepared perovskites [105-110]. Atta et al confirmed the composition of the prepared Sr$_2$PdO$_3$ perovskite and identified the presence of gold nanoparticles over the perovskite surface, Sr$_2$PdO$_3$/Au$_{nano}$, using XPS data [11, 20, 26]. A detailed examination of the oxidation states of A and B-site metal ions using XPS analysis was conducted as a part of structural characterizations of Sr$_2$PdO$_3$ and Sr$_2$PdO$_3$/Au$_{nano}$. Fig. 4.13 (A) revealed the XPS spectra of Sr(3d), Pd(3d) and O(1s) in Sr$_2$PdO$_3$ while Fig. 4.13 (B) revealed the XPS spectra of Sr(3d), Pd(3d), O(1s) and Au(4f) in Sr$_2$PdO$_3$/Au$_{nano}$.

(A) Sr₂PdO₃   (B) Sr₂PdO₃/Au_nano

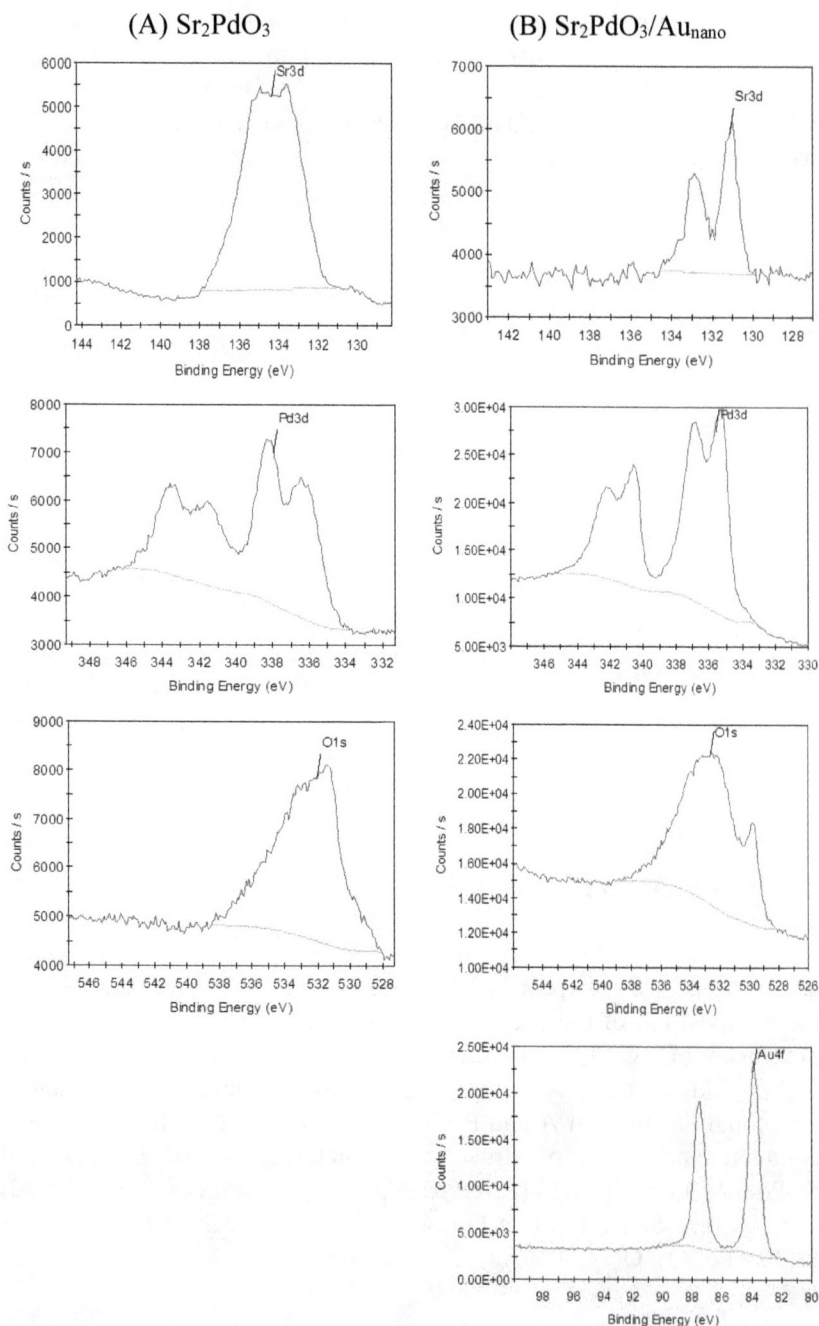

Fig. 4.13. XPS of Sr3d, Pd3d and O1s and Sr3d, Pd3d, O1s and Au4f for **(A)** Sr₂PdO₃ and **(B)** Sr₂PdO₃/Au_nano, respectively, reused with permission [26].

The XPS of Sr3d in Sr_2PdO_3 contained a doublet whose binding energies were 134.4 and 133.2 eV which can be assigned to $Sr3d_{3/2}$ and $Sr3d_{5/2}$ lines, respectively. In case of Sr_2PdO_3/Au_{nano}, they slightly shifted to 133 and 131.1 eV, respectively. The values of the binding energy of Sr(3d) in both samples corresponded to Sr^{2+} ions in perovskite oxide form [105]. In addition, the binding energy values of Sr(3d) in the case of Sr_2PdO_3 and Sr_2PdO_3/Au_{nano} represented Sr^{2+} ions in the form of Sr_2CuO_3 which matched the structure of Sr_2PdO_3 [35] according to the explanation derived by the NIST XPS standard reference database. A slight shift in the binding energy value was observed in case of Sr_2PdO_3 from that of the standard binding energy value assigned for Sr_2CuO_3 by the NIST database. The binding energies of Pd(3d) were 341.6 and 337.94 eV corresponding to $3d_{3/2}$ and $3d_{5/2}$, respectively in case of Sr_2PdO_3 and 342.3 and 335.64 eV, respectively in case of Sr_2PdO_3/Au_{nano}. The metallic Pd peak which was located at 335.2 eV was absent for both samples which confirmed the inclusion of Pd in the perovskite crystal lattice [24, 111]. The values of binding energy of Pd(3d) in both samples revealed the presence of Pd^{2+} in the oxide form corresponding to Sr_2PdO_3 as primary phase and $SrPd_3O_4$ as secondary phase. The peak positions were nearly the same for Sr(3d) and Pd(3d) in both cases revealing that the chemical environment around Sr and Pd didn't change between the two cases. This result confirmed that the deposition of gold nanoparticles didn't affect the chemical structure of Sr_2PdO_3 lattice.

On the other hand, the binding energy of O(1s) was 532.04 eV and a broad shoulder peak at 533.40 eV in case of Sr_2PdO_3 and 532.64 eV and a broad shoulder peak at 533.70 eV in case of Sr_2PdO_3/Au_{nano}. The peak with a low binding energy in both samples (532.04, 532.64 eV) might be attributed to surface lattice oxygen species in the form of O^{2-} ion of the metal oxide [23]. While the other one with a high binding energy (533.40, 533.70 eV) might be assigned to adsorbed oxygen species which might exist in oxygen vacancies of such defect oxides [11, 20, 105]. The two peaks in both cases indicated that the oxygen was bonded in more than one formula with Sr and Pd confirming the presence of primary Sr_2PdO_3 and secondary $SrPd_3O_4$ phases of perovskite. Also, the peak positions were nearly the same in both cases of Sr_2PdO_3 and Sr_2PdO_3/Au_{nano} reflecting the same chemical environment and confirming that the presence of gold nanoparticles didn't affect the chemical structure of Sr_2PdO_3 perovskite. Furthermore, this observation proved the deposition of gold nanoparticles over the surface of perovskite. Furthermore, the binding energy of Au(4f) was 87.90 eV in case of Sr_2PdO_3/Au_{nano} which was ascribed to metallic monovalent Au.

The binding energy of metallic Au is 87.94 eV which was nearly the same as the observed value suggesting that metallic gold nanoparticles were deposited over the surface of Sr_2PdO_3 and didn't affect the structure of the perovskite lattice [26].

4.8. Applications of Perovskites

Inorganic perovskite-type oxides are attractive nano-materials displaying unique physical and chemical properties like super-magnetic, photocatalytic, thermoelectric and dielectric properties, electrically active structure, mobility of oxide ions through the crystal lattice, oxygen content variations, electronic conductivity and thermal and chemical stability. They have a potential role in various applications like solid state chemistry, physics, catalysis, fuel cells, magnetic materials, advanced electrode materials, chemical sensing, electrochemical sensing, etc. This type of nano-oxide exhibits higher catalytic activity than that reported for many transition metal compounds and some precious metal oxides.

Recently, perovskite oxides have opened up new era for possible applications in sensors, particularly electrochemical sensors. Nano-perovskites were utilized in electrochemical sensing of alcohols, gases, amino acids, acetone, glucose, H_2O_2 and neurotransmitters demonstrating fascinating characteristics like unique long term stability, good selectivity, excellent reproducibility, sensitivity and anti-interference ability.

On the other hand, perovskite oxides can be widely used as catalysts in modern chemical industry exhibiting appropriate solid state, surface and morphology properties [6]. Many perovskites show good characteristics like lower activation energy, higher electron transfer rate and higher electrocatalytic activity in the catalysis of oxygen reduction and hydrogen evolution reactions [88]. Also, they have potential roles in the development of anodic catalysts with high catalytic performance for direct fuel cells. Perovskite oxides exhibited fascinating properties like thermal and chemical stabilities, good electrical conductivity similar to that of metals, high ionic conductivity and perfect mixed ionic and electronic conductivity. Depending on the differences in the electrical conductive characteristics of perovskites, they were chosen as an effective component in fuel cells [9]. On the other hand, perovskites are widely applied in many technological areas like electrochromic, photochromic and image storage devices. Also, they are utilized in other device applications like switching devices, infrared detectors, filtering and surface acoustic wave signal processing devices [88].

4.8.1. Gas Sensor

There are a number of requirements that should be realized in a given material to use it as gas sensors: good resemblance with the target gases, manufacturability, hydrothermal stability, convenient electronic structure, resistance to poisoning and adaptation with existing technologies. A wide variety of materials can be applied as gas sensors such as semiconductors: SnO_2, In_2O_3 and WO_3 and perovskite oxides: $LaFeO_3$ and $SrTiO_3$. Perovskite oxides exhibit good characteristics to be applied as gas sensors like catalytic properties, ideal band gap "3–4 eV" and perfect thermal stability. Moreover, the difference in size between the A and B-site cations allows the addition of different dopants for controlling semi-conducting properties. Titanates, ferrites and cobaltates based perovskites are widely used as gas sensors for the detection of different gases like CO, NO_2, methanol, ethanol and hydrocarbons [112-115].

Houshang Alamdari et al prepared a series of $La_{1-x}Ce_xCoO_3$ perovskite-type (x ranging from 0 to 0.2) to be applied as CO gas sensor. The crystallite size of the prepared nanostructured material is around 10 nm and the specific surface area is up to 55 m^2/g. The Ce-doping enhances the surface oxygen and facilitates the adsorption and oxidation processes. The highest conductivity and the lowest activation energy were achieved at $La_{0.9}Ce_{0.1}CoO_3$. A maximum response ratio of 240 % with respect to 100 ppm CO in air was achieved at Ce-doped perovskite compared to 60% obtained with pure $LaCoO_3$. In addition, the optimum CO sensing temperature was found to be 100 °C at the doped sample compared to 130 °C at the pure perovskite [113].

$LaCoO_3$ [112], $NdFe_{1-x}Co_xO_3$ [114] and $La_{0.8}Pb_{0.2}Fe_{0.8}Cu_{0.2}O_3$ [116] were used for CO sensing while $La_{0.8}Pb_{0.2}FeO_3$ for methanol sensing [117], $LaFeO_3$ and its doped forms [118, 119], $SrFeO_3$ [120], $SmFe_{0.9}Mg_{0.1}O_3$ [121], $LaMnO_3$ [122] and other perovskites [123-127] for ethanol sensing, $LaFeO_3$ and $SmFeO_3$ [128-130] for NO_x sensing and $SrTi_{1-x}Fe_xO_{3-\delta}$ [131] and $LnFeO_3$ [132] for hydrocarbons. A summary of various perovskite oxides for different gas sensing was given in Table 4.3.

Table 4.3. A summary of different perovskites for gas sensing.

Perovskite	Sensing for	Oxide type
$LaCoO_3$ [112]	CO	p-type
$La_{0.9}Ce_{0.1}CoO_3$ [113]	CO	p-type
$NdFe_{1-x}Co_xO_3$ [114]	CO	When x<0.3; p-type conduction behavior, when x > 0.3; n-type.
$La_{0.8}Pb_{0.2}Fe_{0.8}Cu_{0.2}O_3$ [116]	CO	p-type
$YCoO_3$ [133]	CO, NO2, NO, and CH4	p-type
$GdInO_3$ [134]	CO	p-type
$LaFeO_3$ [135]	liquefied petroleum gas	p-type
$La_{0.8}Pb_{0.2}FeO_3$ [117]	Methanol	p-type
$LaFeO_3$ [118]	Ethanol	p-type
$LaMg_{0.1}Fe_{0.9}O_3$ [119]	Ethanol	p-type
$SrFeO_3$ [120]	Ethanol	p-type
$SmFe_{0.9}Mg_{0.1}O_3$ [121]	Ethanol	p-type
$LaMnO_3$ [122]	Ethanol	Not reported
$La_{0.875}Ba_{0.125}FeO_3$ [123]	Ethanol	p-type
$Ca_xLa_{1-x}FeO_3$ [124]	Ethanol	Not reported
$LaCo_{0.1}Fe_{0.9}O_3$ [125]	Ethanol	p-type
$LaFeO_3$ [126]	Ethanol	p-type
$La_{1-x}Pb_xFeO_3$ [127]	Ethanol	x = 0.30; p-type, x = 0.32–0.50; n-type
$LaFeO_3$ and $SmFeO_3$ [128]	NO2, CO	p-type
$LaFeO_3$ [129]	NO2	Not reported
$LaFeO_3$ [130]	NOx	Not reported
$SrTi_{1-x}Fe_xO_{3-\delta}$ [131]	hydrocarbons	p-type
$LnFeO_3$ (Ln = La, Nd and Sm) [132]	hydrocarbons	p-type

4.8.2. Glucose and H_2O_2 Sensor

The analytical determination of H_2O_2 and glucose is very crucial in many areas such as food, clinic and pharmaceutical analyses. One of the most important oxidizing agents in chemical and food industries is H_2O_2. Glucose is considered a main metabolite for living organisms and for diabetes mellitus clinical examination. Thus, it is very necessary to sensitively and selectively determine H_2O_2 and glucose using appropriate sensors [136-145]. Several types of enzymatic glucose sensors were mentioned in literature exhibiting the characteristics of simplicity and sensitivity. However, enzymatic glucose sensors have some drawbacks like enzyme stability lack (due to the intrinsic its intrinsic nature of enzyme as its activity was highly affected by poisonous chemicals, pH, temperature, humidity, etc.) and difficult steps needed for the enzyme immobilization on the electrode surface. Therefore, sensitive, stable and selective non-enzymatic glucose sensors have been constructed and different novel materials have been used for the electrocatalytic oxidation of glucose like noble nano-metals, nano-alloys, metal oxides and inorganic perovskite oxides. Perovskites have been used as glucose sensors due to their fascinating properties like superconductivity, catalytic activity, good biocompatibility, ferroelectricity, charge ordering, high thermo-power and the ability perovskite structure to accept different metallic ions in their lattice [136-145].

Atta et al studied the modification of Sr_2PdO_3 perovskite modified graphite electrode with gold nanoparticles, graphite/Sr_2PdO_3/Au_{nano}, and examined its catalytic activity toward the enzymatic-free glucose sensing. Sr_2PdO_3 perovskite was prepared by glycine-nitrate combustion method. Then gold nanoparticles were deposited electrochemically on the surface of graphite/Sr_2PdO_3. The optimized nanocomposite would be utilized in the sensing applications as non-enzymatic voltammetric glucose sensor. The electrochemical characterization of this study explored the high synergism achieved by Sr_2PdO_3 and Au_{nano} for glucose electrocatalysis (Fig. 4.14). The utilized nanocomposite offered many advantages for glucose electro-oxidation such as high sensitivity, perfect selectivity, anti-interference ability, low detection limit and excellent long term stability. Moreover, this non-enzymatic glucose sensor was highly selective to glucose even in presence of common interferences like ascorbic acid, uric acid, paracetamol, dopamine and chloride. Furthermore, this free-enzyme nanocomposite was utilized for glucose sensing in human urine samples with excellent recovery results and good reproducibility [26, 145].

In addition, Atta et al examined the effect of doping the B-site "Pd^{2+}" of Sr_2PdO_3 with another metal "Au^{3+}" and investigated the performance of perovskite-type oxides $Sr_2Pd_{1-y}Au_yO_3$ with various substitution degrees (y) with the aim of developing a highly sensitive and selective non-enzymatic glucose sensor. A series of nano-metric $Sr_2Pd_{1-y}Au_yO_3$ perovskites were prepared demonstrating high electrocatalytic activity for non-enzymatic glucose sensing. Doping the Sr_2PdO_3 perovskite with Au^{3+}, which has higher conductivity than Pd^{2+}, improved both the electrocatalytic activity and the electrical conductivity compared to the parent material Sr_2PdO_3. $Sr_2Pd_{0.7}Au_{0.3}O_3$, with the optimum ratio of Au dopant, showed higher electrocatalytic activity toward non-enzymatic glucose sensing compared to the other ratios.

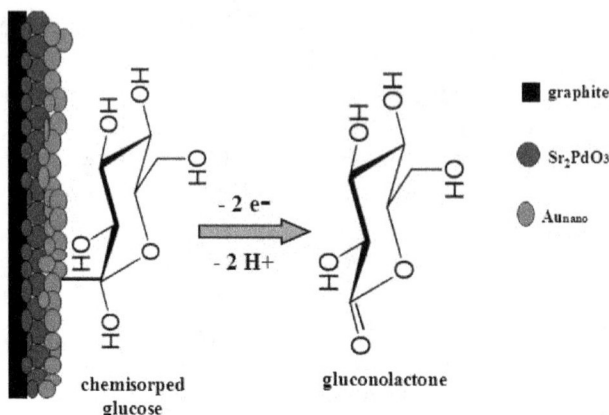

Fig. 4.14. Mechanism of oxidation of glucose at the proposed sensor, graphite/Sr_2PdO_3/Au_{nano}, reused with permission [145].

The proposed non-enzymatic glucose sensor, $Sr_2Pd_{0.7}Au_{0.3}O_3$, offered high performance in terms of good selectivity in presence of common interferents, noticeable sensitivity, low detection limit, wide linearity, excellent long term stability, anti-interference ability and applicability in real sample analysis [26].

On the other hand, Atta et al examined the effect of the partial substitution of the Sr^{2+} A-site in Sr_2PdO_3 by Ca^{2+} ions ($Sr_{2-x}Ca_xPdO_3$ with x= 0–0.7) on the structural and catalytic properties, particularly the electrocatalytic activity in non-enzymatic glucose sensing. $Sr_{2-x}Ca_xPdO_3$ with various substitution degrees (x) was prepared and the preparation conditions and the amount and type of dopant were optimized. Well-crystalline orthorhombic Sr_2PdO_3 phase was formed as the main phase in all the prepared samples with particles size in the range of nanometers as confirmed by XRD. Fig. 4.15 (A) showed the CVs of 5 mmol L^{-1} glucose/0.1 mol L^{-1} NaOH at bare graphite, graphite/Sr_2PdO_3 and graphite/$Sr_{1.7}Ca_{0.3}PdO_3$. Higher current response was obtained at graphite/$Sr_{1.7}Ca_{0.3}PdO_3$ compared to other electrodes. Doping the A-site Sr^{2+} cations with Ca^{2+} ions in Sr_2PdO_3 perovskite led to the enhancement of the elecrocatalytic activity towards non-enzymatic glucose sensing showing the highest elecrocatalytic activity in case of $Sr_{1.7}Ca_{0.3}PdO_3$. This might be attributed to the higher free volume in the crystal lattice, enhanced content and mobility of surface lattice oxygen and distorted stabilized perovskite structure leading to higher catalytic activity of the prepared perovskites. In addition, synergistic interactions were achieved between Sr^{2+} and Ca^{2+} in the A-site and Pd^{2+} ions in the B-site resulted in improved surface activity and stabilized structure. As a result, greater ionic and electronic conductivity and enhanced catalytic activity were achieved upon doping compared to the parent Sr_2PdO_3 [26, 34].

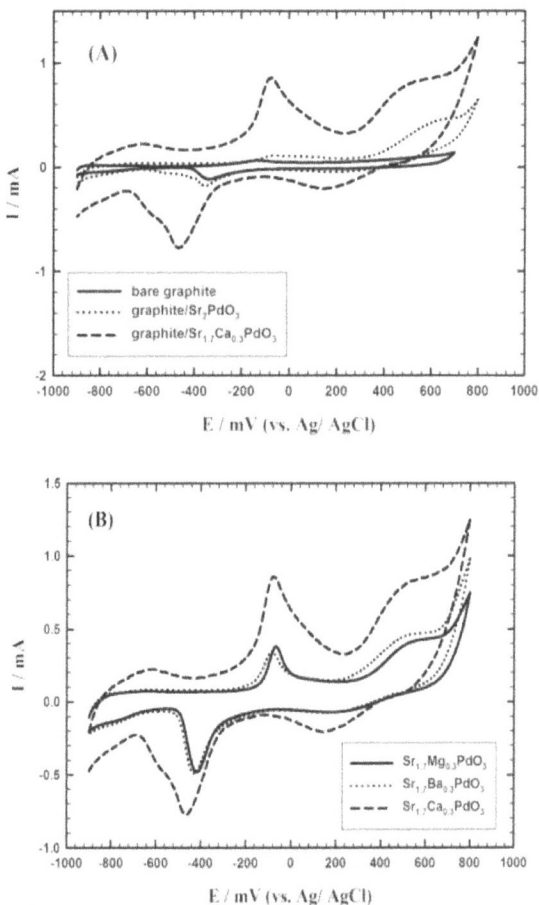

Fig. 4.15. (A) CVs of 5 mmol L^{-1} glucose/0.1 mol L^{-1} NaOH at bare graphite, graphite/Sr_2PdO_3 and graphite/$Sr_{1.7}Ca_{0.3}PdO_3$, scan rate 50 mV s^{-1}. **(B)** CVs of 5 mmol L^{-1} glucose/0.1 M NaOH at graphite/$Sr_{1.7}M_{0.3}PdO_3$ (M: Mg, Ca and Ba), scan rate 50 mV s^{-1}, reused with permission [34].

Also, the type of dopant used was studied and the CVs of 5 mmol L^{-1} glucose/0.1 mol L^{-1} NaOH at graphite/$Sr_{1.7}M_{0.3}PdO_3$ (M; Mg, Ca, Ba) was shown in Fig. 4.15 (B). Higher current response was obtained in case of Ca^{2+} dopant compared to other types. In addition, utilization of Ca^{2+} ions as a dopant in Sr_2PdO_3 showed greater conductivity and higher electrocatalytic activity compared to other dopants (Mg^{2+}, Ba^{2+}). Ca^{2+} dopant enhanced the electronic properties of Sr_2PdO_3 perovskite and facilitated the electron transfer rate. Graphite/$Sr_{1.7}Ca_{0.3}PdO_3$ as a free enzymatic glucose sensor exhibited good stability, low detection limit, high sensitivity, good selectivity even in presence of common interferents, applicability in real sample analysis and anti-interference ability [26, 34].

Different types of perovskites were used for enzymatic and non-enzymatic H_2O_2 and glucose sensing exhibiting high sensitivity, wide linear range, low detection limit, anti-interference ability, applicability in real samples and long term stability. A brief summary of various types of perovskites is given in Table 4.4.

4.8.3. Neurotransmitters Sensor

Dopamine "DA" is one of the most important catecholamines present in the mammalian central nervous system. DA depletion leads to Parkinson's disease thus it is very important to determine its concentration. The major problem in DA detection is the interference of ascorbic acid "AA" and uric acid "UA" with it [91, 151-154]. It is very crucial to construct a sensor that can be sensitively and selectively detect DA even in presence of high concentration of AA and UA.

Atta et al fabricated a novel sensor for dopamine based on the modification of carbon paste electrode with Sr_2PdO_3 (CpE/Sr_2PdO_3). Higher current response, faster electron transfer kinetics and lower overpotential toward DA sensing were obtained at CpE/Sr_2PdO_3 compared to palladium nanoparticles modified electrode and bare CpE. This is attributed to the oxygen–surface interaction that was achieved between the oxygen atoms of the hydroxyl groups of DA and the transition element in the perovskite. The perovskite shows a deficiency of its surface for oxygen thus a "moderate" bond between oxygen atoms and the transition element in the oxide is formed based on the adsorption of "dihydroxy"-oxygen onto perovskite surface "Fig. 4.16" [91]. The proposed sensor exhibits long term stability, low detection limit of 9.3 nmol L^{-1} and anti-interference ability even in presence of high concentrations of AA and UA [26, 91].

Table 4.4. A summary of different perovskites for H_2O_2 and glucose sensing.

Perovskite	Sensing for	Sensor type	Detection limit	Sensitivity
LaTiO$_3$-Ag$_{0.1}$ [136]	glucose	Non-enzymatic	2.50×10^{-9} M	7.80×10^2 μA mM^{-1}cm^{-2}
LaNi$_{0.5}$Ti$_{0.5}$O$_3$/CoFe$_2$O$_4$ [137]	H_2O_2	Non-enzymatic	23 nM	3.21 μA μM^{-1} cm^{-2}
LaNi$_{0.6}$Co$_{0.4}$O$_3$ [138]	H_2O_2 and glucose	Non-enzymatic	1 nM H_2O_2 8 nM glucose	1812.84 μA mM^{-1} cm^{-2} for H_2O_2 and 643.0 μA mM^{-1} cm^{-2} for glucose
CPE–La$_{0.66}$Sr$_{0.33}$MnO$_3$–GO$_x$ (glucose oxidase) [139]	glucose	Enzymatic	Not reported	158.1 μAmol^{-1} L
Co$_{0.4}$Fe$_{0.6}$LaO$_3$ [140]	H_2O_2 and glucose	Non-enzymatic	2 nM H_2O_2 and 10 nM glucose	Not reported
LaTiO$_3$–Ag$_{0.2}$ [141]	glucose	Non-enzymatic	2.1×10^{-7} M	784.14 μA mM^{-1} cm^{-2}
LaNi$_{0.5}$Ti$_{0.5}$O$_3$–NiFe$_2$O$_4$ [142]	glucose	Enzymatic	0.04 mM	Not reported
La$_{0.6}$Ca$_{0.4}$Ni$_{0.7}$Fe$_{0.3}$O$_3$ [143]	H_2O_2	Non-enzymatic	Not reported	Not reported
La$_{0.5}$Sr$_{0.5}$CoO$_{3-\delta}$ [144]	H_2O_2	Non-enzymatic	Not reported	Not reported
graphite/SrPdO$_3$/Au$_{nano}$ [145]	glucose	Non-enzymatic	10.1 μmol L^{-1}	422.30 μA/mmol L^{-1}
La$_{0.7}$Sr$_{0.3}$NiO$_3$ /chitosan/GCE [146]	H_2O_2	Enzymatic	9.0×10^{-8} mol/L	Not reported
LaNiO$_3$ [147]	H_2O_2 and glucose	Non-enzymatic	33.9 nM H_2O_2	Not reported
Sr$_2$Pd$_{0.7}$Au$_{0.3}$O$_3$ [26]	glucose	Non-enzymatic	0.498 nmol L^{-1}	14355.2 μA/mmol L^{-1} cm^2
Sr$_{1.7}$Ca$_{0.3}$PdO$_3$ [26, 34]	glucose	Non-enzymatic	0.0845 μmol L^{-1}	306.9 μA/mmol L^{-1}
La$_{0.88}$Sr$_{0.12}$MnO$_3$ [148]	glucose	Non-enzymatic	31.2 nM	1111.1 μA mM^{-1}cm^{-2}
KNbO$_3$ [149]	H_2O_2	Enzymatic	Not reported	750 μA mM^{-1} cm^{-2}
Sr$_{0.85}$Ce$_{0.15}$FeO$_3$ [150]	H_2O_2	Non-enzymatic	10 μM	60 μA mM^{-1} cm^{-2}

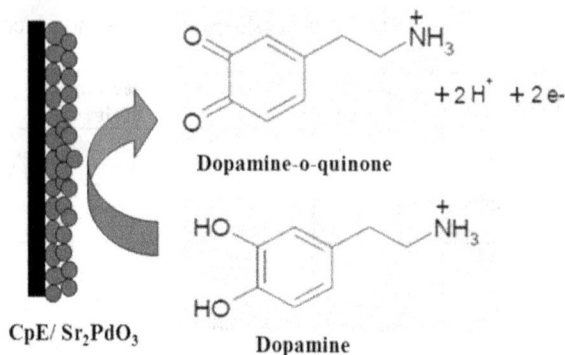

Fig. 4.16. The electrochemical oxidation of DA at CpE/Sr$_2$PdO$_3$, reused with permission [26].

In addition, Atta et al studied the effect of the substrate used for the immobilization of Sr$_2$PdO$_3$ perovskite. Two different substrates: carbon paste (CpE) and graphite electrodes were compared toward the electrocatalytic oxidation of some neurotransmitters and biologically important compounds. The studied compounds are: dopamine (DA), epinephrine (EP), norepinephrine (NE), serotonin (ST), 3,4-dihydroxyphenylacetic acid (DOPAC) and L-dopa (Fig. 4.17). More resolved peak, less oxidation potential and higher current density were obtained for all the studied compounds at graphite/Sr$_2$PdO$_3$ compared to CpE/Sr$_2$PdO$_3$. It was confirmed that graphite electrode was more suitable substrate for immobilization of Sr$_2$PdO$_3$ perovskite than CpE due to higher porosity, larger surface area and greater catalytic activity. Graphite/Sr$_2$PdO$_3$ exhibited higher electrocatalytic activity toward the studied neurotransmitters compared to bare graphite and other well-known neurotransmitters sensors. The difference in the electrocatalytic activity of the proposed electrodes was explained in terms of SEM (Fig. 4.18). The proposed sensor showed many advantages such as lower detection limit (1.63 nmol L^{-1}) and higher sensitivity (0.796 µA/µmol L^{-1}) toward L-dopa sensing. Also, it exhibited longer term stability (up to two months) toward DA sensing. On the other hand, this sensor showed anti-interference ability, it can simultaneously determine L-dopa in presence of a large amount of UA and AA as common interferences in biological fluids. In addition, it can simultaneously discriminate L-dopa from ST in the presence of excess amount of AA. This method was successfully applied for determination of L-dopa in human urine samples with good precision, accuracy, selectivity and very low detection limit [26, 152].

186

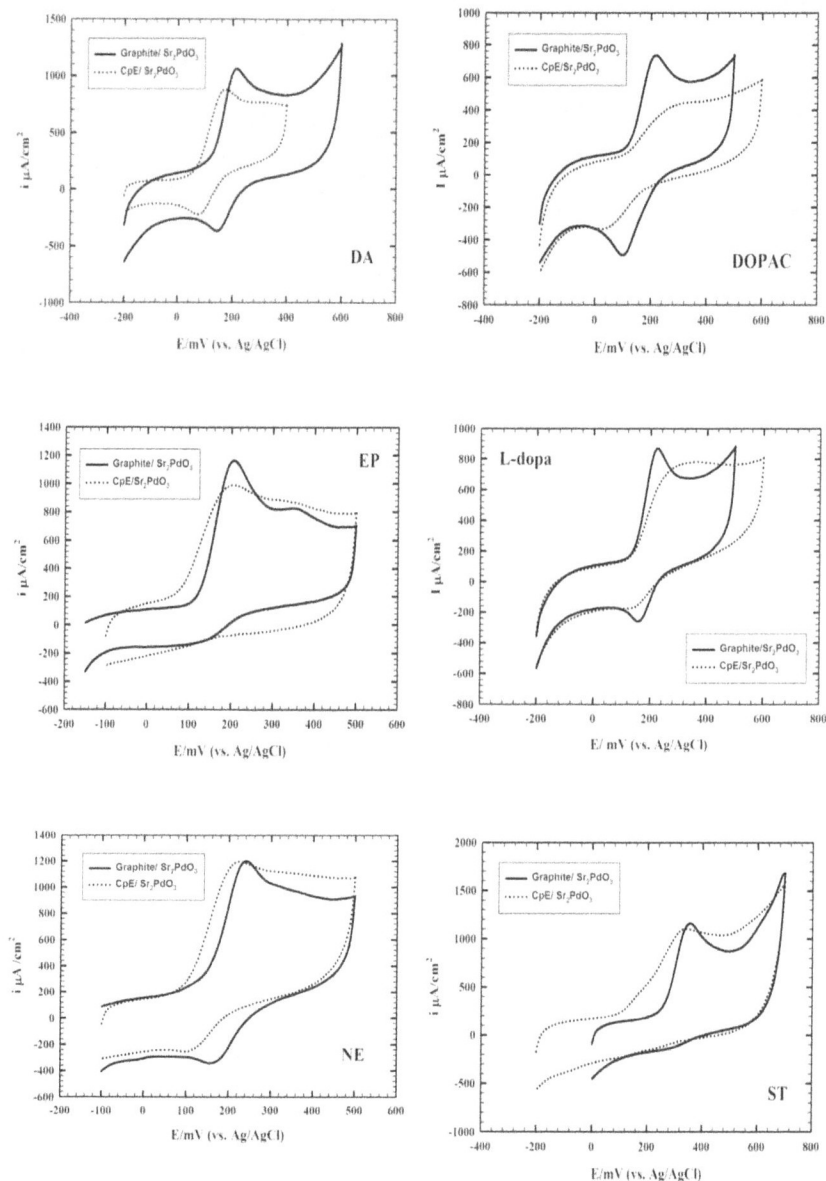

Fig. 4.17. CVs of 1 mmol L^{-1} of each compound (DA, DOPAC, EP, L-dopa, NE and ST)/0.1 mol L^{-1} PBS/pH 7.40 at graphite/Sr$_2$PdO$_3$ (solid line) and CpE/Sr$_2$PdO$_3$ (dash line), scan rate 50 mV s^{-1}, reused with permission [152].

187

(a)   (b)

(c)

Fig. 4.18. SEM micrographs of (a) CPE/Sr_2PdO_3, (b) graphite/Sr_2PdO_3 and (c) greater magnification 100,000×, reused with permission [152].

In addition, $LaFeO_3$ micro-spheres made up of nano-spheres [151], $LaFeO_3$ nanostructure dendrites [153], $LaFeO_3$ nanoparticles [154] and $LaMnO_3$ [155] were used for DA sensing with low detection limit of 59 nM, 62 nM, 30 nM and 6.22 μM, respectively. A summary of different perovskites used for neurotransmitters sensing was given in Table 4.5.

4.8.4. Sensor for Drugs

Nano-perovskites have been widely utilized as sensors due to their fascinating properties like superconductivity, catalytic activity, good biocompatibility, ferroelectricity, charge ordering, high thermo-power and the ability of the perovskite structure to accept different metallic ions in their lattice. Many papers in the literature mentioned the different synthesis routes of many perovskites and their applications in catalysis, fuel cell and gas and glucose sensing and but no publications examined their applications in the electrochemical sensing of pharmaceutical drugs.

Table 4.5. A summary of different perovskites for neurotransmitters sensing.

Perovskite	Sensing for	Linear dynamic range	Detection limit (nM)
LaFeO$_3$ microspheres made up of nanospheres [151]	DA	$2 \times 10^{-8} - 1.6 \times 10^{-6}$ M	59
CpE/Sr$_2$PdO$_3$ [91]	DA	$7 - 70$ µmol L^{-1}	9.3
Graphite/Sr$_2$PdO$_3$ [152]	L-dopa EP, NE, DA, DOPAC and ST	$0.6 - 9$ µmol L^{-1}	1.63
LaFeO$_3$ nanostructure dendrites [153]	DA	$8.2 \times 10^{-8} - 1.6 \times 10^{-7}$ M	62
LaFeO$_3$ nanoparticles [154]	DA	$1.5 \times 10^{-7} - 8.0 \times 10^{-4}$ M	30
LaMnO$_3$ [155]	DA	5-50 µM	6.22 µM

Atta et al presented a sensitive novel approach for the electrochemical determination of ketotifen drug in pharmaceutical formulations for the first time based on the in-situ modification of carbon paste electrode (CpE) with NdFeO$_3$ (NF) in presence of sodium dodecyl sulfate (SDS), NFMCpE...SDS. Ketotifen was pharmaceutically important as it was utilized for the treatment of asthma attacks, rhinitis, skin allergies and anaphylaxis. The ortho-ferrite neodymium nano-crystals (NdFeO$_3$) were synthesized via simple microwave assistant-citrate method. Few publications mentioned the applications of NdFeO$_3$ in catalysis and fuel cells but no papers reported its electrochemical sensing applications. The proposed sensor "NFMCpE...SDS" demonstrated greater electro-catalytic activity, higher sensitivity and better selectivity toward ketotifen oxidation compared to NFMCpE, CpE...SDS and bare CpE. The utilization of surface-active agents in this work area presented a novel useful dimension towards ketotifen investigation. NdFeO$_3$ perovskite and SDS acted as catalytic modifiers facilitating the electron transfer rate of ketotifen at NFMCpE...SDS. In addition, the surface area of the proposed sensor affected greatly its catalytic activity toward ketotifen determination compared to the other electrodes. Moreover, the proposed sensor could effectively detect ketotifen even in presence of common interferents, pain relief drugs, neurotransmitters or narcotics. The proposed sensor showed perfect characteristics such as high sensitivity, repeatability, low detection limit, wide linear range, anti-interference ability, applicability in real samples and long term stability [26].

Furthermore, Atta et al presented a simple electrochemical sensor for metoclopramide drug for the first time based on the in-situ modification of carbon paste electrode (CpE) with $NdFeO_3$ (NF) nano-perovskite and 2-chloro-1,3-dimethyl-imidazolinium hexafluorophosphate ionic liquid crystal (ILC) in the presence of sodium dodecyl sulfate (SDS), NFILCMCpE…SDS. Metoclopramide drug displayed a pharmaceutical importance as an antiemetic and gastroprokinetic drug. This work combined the effect of $NdFeO_3$ in improving the catalytic activity and ILC in enhancing the conductivity with the effect of SDS to facilitate the preconcentartion/accumulation of metoclopramide drug at the electrode surface. Evidence of the formation of a core-shell like structure between the nano-perovskite particles and the ILCs was demonstrated (Fig. 4.19).

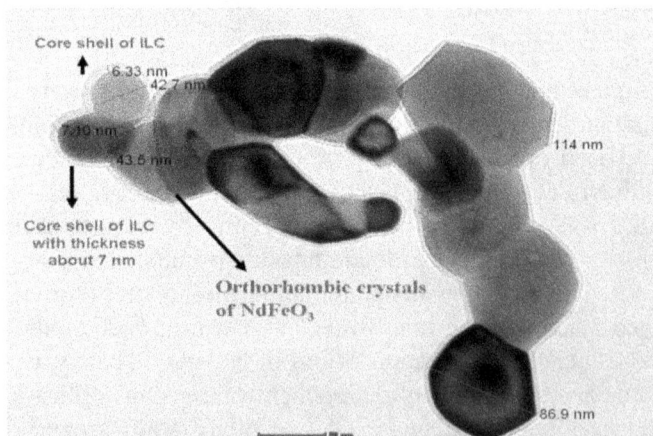

Fig. 4.19. TEM micrograph of $NdFeO_3$/ILC, reused with permission [92].

Cyclic voltammetry and electrochemical impedance spectroscopy measurements revealed enhanced electrocatalytic activity for metoclopramide at the proposed sensor. The catalytically active $NdFeO_3$, super anisotropic ionic conductive and highly polarizable ILC and the preconcentrating agent SDS explored synergistic effect toward metoclopramide determination resulting in facilitation of charge transfer rate (Fig. 4.20).

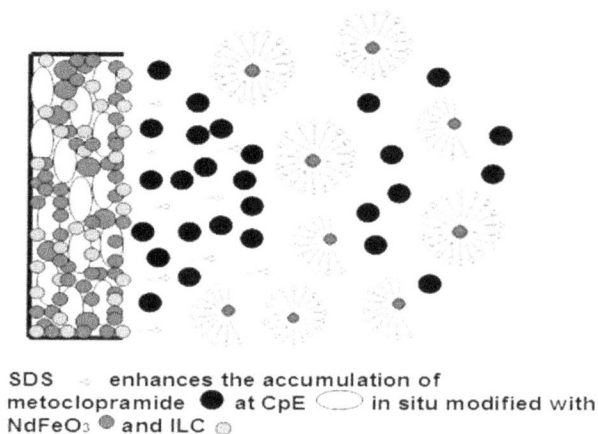

SDS enhances the accumulation of metoclopramide ● at CpE ⊂⊃ in situ modified with NdFeO₃ ● and ILC ○

Fig. 4.20. Schematic diagram of accumulation of metoclopramide at NFILCMCpE...SDS, reused with permission [92].

Also, the greater electrocatalytic activity of the proposed sensor was related to the higher real active surface area. The BET surface area was 26.81 m^2/g in case of NdFeO$_3$ perovskite which increased to 44.51 m^2/g in case of NdFeO$_3$/ILC. In addition, ILC exhibited superior feature of spontaneous orientation which might promote NdFeO$_3$ inside the paste to form highly ordered molecularly packed microstructure leading to dramatic increase in the conductivity of the proposed sensor. Based on the electrocatalytic effect of the proposed sensor, DPV was developed for the determination of metoclopramide in the linear range of 0.4 μmol L^{-1} to 300 μmol L^{-1} with very low detection limit of 1.91 nmol L^{-1} and high sensitivity of 203.56 μA/mmol L^{-1}. Very low detection limit of 1.73 nmol L^{-1} and 3.68 nmol L^{-1} was achieved in urine samples and blood serum, respectively. Also, the proposed sensor exhibited anti-interference properties toward metoclopramide even in presence of common interferents, pain relief drugs or neurotransmitters. The proposed sensor was very stable toward the electrochemical response of metoclopramide up to one month of storage. The proposed sensor displayed good performance in terms of selectivity, wide linearity range, sensitivity, low detection limit, applicability in real samples of urine, blood and pharmaceutical tablets, long term stability and anti-interference ability. The proposed sensor opened up a new era of research and applications for sensing several and different number of pharmaceutical drugs with high sensitivity and selectivity [26, 92]. A summary of different perovskites used for drugs sensing was given in Table 4.6.

Table 4.6. A summary of different perovskites for drugs sensing.

Perovskite	Sensing for	Linear dynamic range	Detection limit (nM)
NdFeO$_3$ [26]	Ketotifen	0.8 µM to 360 µM	2.92
NdFeO$_3$ [26, 92]	Metoclopramide	0.4 µM to 300 µM	1.91
KSr$_2$Ni$_{0.75}$Nb$_{4.25}$O$_{15-\delta}$ [156]	Dipyrone	3.5×10^{-5} - 3.1×10^{-4} M	5.1 µM

4.8.5. Sensor for Amino Acids

Amino acids like L-cysteine (L-cys), L-tryptophane (L-trp), L-alanine (L-ala) and L-phenylalanine (L-phe) are considered as important building blocks of biological compounds. Also, they play very essential role in many biochemical processes. Perovskite modified electrodes present good electrocatalytic activity toward the oxidation of amino acids due to their unique characteristics. LaNi$_{0.5}$Ti$_{0.5}$O$_3$ modified carbon paste electrode was utilized as a novel sensor for amino acids. Under the optimized conditions, LaNi$_{0.5}$Ti$_{0.5}$O$_3$ showed good reproducibility, long-term stability and fast current response toward the determination of different amino acids (L-cys, L-trp, L-ala and L-phe). The proposed material showed catalytic performance toward the determination of amino acids in terms of detection limit, sensitivity and linear dynamic range [157]. The detection limit was 0.5 µM, 0.67 µM, 8 µM and 3 µM for L-cys, L-trp, L-ala and L-phe, respectively [157]. In addition, Liang Bian et al constructed a novel sensor for amino acids based on BiFeO$_3$ exhibiting good performance due to its electrical sensitivity [158].

References

[1]. T. Wolfram, S. Ellialtioğlu, Electronic and Optical Properties of d-band Perovskites, 1st ed. *Cambridge University Press,* New York, 1, 2006.

[2]. F. S. Galasso, Structure, properties and preparation of perovskite-type compounds, in Smoluchowski R., Kurti N., (Eds.), 1st ed., *Pergamon Press,* New York, Chapter 2, 1969, pp. 3-49.

[3]. S. M. Ali, Synthesis of Nano-particles Using Microwave Technique, the Study of their Physical Properties and Some Applications, PhD Thesis, Faculty of Science, *Cairo University,* 2009.

[4]. Mark Levy, Crystal Structure and Defect Properties in Ceramic Materials, Chapter 3: Perovskite Perfect Lattice, PhD Theses, *Atomistic Simulation Group,* 2005, pp. 79-114.

[5]. M. Johnsson, P. Lemmens, Crystallography and Chemistry of Perovskites, in Handbook of Magnetism and Advanced Magnetic Materials, *Wiley,* 2005, pp. 1-11.

[6]. M. A. Peňa, J. L. G. Fierro, Chemical Structures and Performance of Perovskite Oxides, *Chemical Reviews,* Vol. 101, 2001, pp. 1981-2017.

[7]. J. C. Rendón-Angeles, Z. Matamoros-Veloza, K. L. Montoya-Cisneros, J. L. Cuevas, K. Yanagisawa, Synthesis of Perovskite Oxides by Hydrothermal Processing – From Thermodynamic Modelling to Practical Processing Approaches, *InTech,* Chapter 2, 2016, pp. 27-81.

[8]. P. Porta, S. De Rossi, M. Faticanti, G. Minelli, I. Pettiti, L. Lisi, M. Turco, Perovskite-Type Oxides: I. Structural, Magnetic, and Morphological Properties of LaMn1-xCuxO3 and LaCo1-xCuxO3 Solid Solutions with Large Surface Area, *Journal of Solid State Chemistry,* Vol. 146, 1999, pp. 291-304.

[9]. T. Ishihara, Perovskite Oxide for Solid Oxide Fuel Cells, in Fuel Cells and Hydrogen Energy, Ishihara T. (Ed.), *Springer,* Chapter 1, 2009, pp. 1-16.

[10]. A. D. Lozano-Gorrín, Structural Characterization of New Perovskites, in Polycrystalline Materials–Theoretical and Practical Aspects, Zakhariev Z., (Ed.), *InTech,* Chapter 5, 107-124, 2012.

[11]. Z. Li, M. Meng, Q. Li, Y. Xie, T. Hu, J. Zhang, Fe-substituted nanometric La0.9K0.1Co1-xFexO3-δ perovskite catalysts used for soot combustion, NOx storage and simultaneous catalytic removal of soot and NOx, *Chemical Engineering Journal,* Vol. 164, 2010, pp. 98-105.

[12]. G. R. O. Silva, J. C. Santos, D. M. H. Martinelli, A. M. G. Pedrosa, M. J. B. Souza, D. M. A. Melo, Synthesis and Characterization of LaNixCo1-xO3 Perovskites via Complex Precursor Methods, *Materials Sciences and Applications,* Vol. 1, 2010, pp. 39-45.

[13]. A. Jahangiri, H. Aghabozorg, H. Pahlavanzadeh, Effects of Fe substitutions by Ni in La–Ni–O perovskite-type oxides in reforming of methane with CO_2 and O_2, *International Journal of Hydrogen Energy,* Vol. 38, 2013, pp. 10407-10416.

[14]. N. Russo, P. Palmisano, D. Fino, Pd substitution effects on perovskite catalyst activity for methane emission control, *Chemical Engineering Journal,* Vol. 154, 2009, pp. 137-141.

[15]. M. Oishi, K. Yashiro, K. Sato, J. Mizusaki, T. Kawad, Oxygen nonstoichiometry and defect structure analysis of B-site mixed perovskite-type oxide (La, Sr)(Cr, M)O₃−δ (M=Ti, Mn and Fe), *Journal of Solid State Chemistry,* Vol. 181, 2008, pp. 3177-3184.

[16]. X. Liu, H. Zhao, J. Yang, Y. Li, T. Chen, X. Lu, W. Ding, F. Li, Lattice characteristics, structure stability and oxygen permeability of BaFe1−xYxO₃−δ ceramic membranes, *Journal of Membrane Science,* Vol. 383, 2011, pp. 235-240.

[17]. J. Dho, N. H. Hur, Magnetic and transport properties of lanthanum perovskites with B-site half doping, *Solid State Communications,* Vol. 138, 2006, pp. 152-156.

[18]. J. Suntivich, K. J. May, H. A. Gasteiger, J. B. Goodenough, Y. Shao-Horn, A perovskite oxide optimized for oxygen evolution catalysis from molecular orbital principles, *Science,* Vol. 334, 2011, pp. 1383-1385.

[19]. A. Vojvodic, J. K. Nørskov, Optimizing Perovskites for the Water-Splitting Reaction, *Science,* Vol. 334, 2011, pp. 1355-1356.

[20]. M. Alifanti, R. Auer, J. Kirchnerova, F. Thyrion, P. Grange, B. Delmon, Activity in methane combustion and sensitivity to sulfur poisoning of La1−xCexMn1−yCoyO₃ perovskite oxides, *Applied Catalysis,* B, Vol, 41, 2003, pp. 71-81.

[21]. T. Kida, A. Yamasaki, K. Watanabe, N. Yamazoe, K. Shimanoe, Oxygen-permeable membranes based on partially B-site substituted BaFe1−yMyO₃−δ (M=Cu or Ni), *Journal of Solid State Chemistry,* Vol. 183, 2010, pp. 2426-2431.

[22]. I. N. Sora, T. Caronna, F. Fontana, C. D. J. Fernàndez, A. Caneschi, M. Green, Crystal structures and magnetic properties of strontium and copper doped lanthanum ferrites, *Journal of Solid State Chemistry,* Vol. 191, 2012, pp. 33-39.

[23]. G. C. M. Rodríguez, B. Saruhan, Effect of Fe/Co-ratio on the phase composition of Pd-integrated perovskites and its H2-SCR of NOx performance, *Applied Catalysis,* B, Vol. 93, 2010, pp. 304-313.

[24]. Z. Ke-bin, C. Hong-de, T. Qun, Z. Bao-wei, S. Di-xin, X. Xiao-bai, Synergistic effects of palladium and oxygen vacancies in the Pd/perovskite catalysts synthesized by the spc method, *Journal of Environmental Sciences,* Vol. 17, 1, 2005, pp. 19-24.

[25]. C. Li, B. Jiang, W. Fanchiang, Y. Lin, The effect of Pd content in LaMnO₃ for methanol partial oxidation, *Catalysis Communications,* Vol. 16, 2011, pp. 165-169.

[26]. E. H. El-Ads, Characterization and some sensing applications of nano-inorganic oxides [PhD thesis], Faculty of Science, *Cairo University,* 2016.

[27]. F. W. Poulsen, Defect chemistry modelling of oxygen-stoichiometry, vacancy concentrations, and conductivity of (La1−xSrx)yMnO3±δ, *Solid State Ionics,* Vol. 129, 2000, pp. 145-162.

[28]. A. Parveen, N. K. Gaur, Effect of A-site doping on thermal properties of LaGaO3, *Solid State Sciences*, Vol. 14, 2012, pp. 814-819.

[29]. A. Srivastava, N. K. Gaur, N. Kaur, R. K. Singh, Effect of cation doping on low-temperature specific heat of $LaMnO_3$ manganite, *Journal of Magnetism and Magnetic Materials*, Vol. 320, 2008, pp. 2596-2601.

[30]. L. Malavasi, C. Ritter, M. C. Mozzati, C. Tealdi, M. S. Islam, C. B. Azzoni, G. Flor, Effects of cation vacancy distribution in doped $LaMnO_3+\delta$ perovskites, *Journal of Solid State Chemistry*, Vol. 178, 2005, pp. 2042-2049.

[31]. F. Lin, W. Shi, Effects of doping site and pre-sintering time on microstructure and magnetic properties of Fe-doped $BaTiO_3$ ceramics, *Physica B*, Vol. 407, 2012, pp. 451-456.

[32]. H. J. Wei, Y. Cao, W. J. Ji, C. T. Au, Thermally stable ceria–zirconia catalysts for soot oxidation by O_2, *Catalysis Communications*, Vol. 9, 2008, pp. 250-255.

[33]. Toshiro Kohmoto, Doping-Induced Ferroelectric Phase Transition and Ultraviolet-Illumination Effect in a Quantum Paraelectric Material Studied by Coherent Phonon Spectroscopy, in Handbook of Advances in Ferroelectrics, *InTech*, Chapter 12, 2013, pp. 257-278.

[34]. Ekram H. El-Ads, Ahmed Galal and Nada F. Atta, The effect of A-site doping in a strontium palladium perovskite and its applications for non-enzymatic glucose sensing, *RSC Advances*, Vol. 6, 2016, pp. 16183-16196.

[35]. O. Miller and R. Roy, Synthesis and Crystal Chemistry of Some New Complex Palladium Oxides, Advances in Chemistry, Robert F. Gould (Ed.), 158th Meeting of the American Chemical Society, New York, N. Y., September 8-9, 1969.

[36]. Y. Tsujimoto, K. Yamaura and E. Takayama-Muromachi, Oxyfluoride Chemistry of Layered Perovskite Compounds, *Applied Science*, Vol. 2, 2012, pp. 206-219.

[37]. P. Villars, Springer & Material Phases Data System (MPDS), CH-6354 Vitznau, Switzerland & National Institute for Materials Science (NIMS), Springer Materials Sr_2PdO_3 Crystal Structure, Japan, 2014, Database name LINUS PAULING FILE Multinaries Edition – 2012, Dataset ID sd_0375642.

[38]. P. Villars, K. Cenzual and R. Gladyshevskii, Hand book of inorganic substances, *De Gruyter*, 2015.

[39]. M. A. Augustyniak-Jabłokowa, I. Jacyna-Onyszkiewiczb, T. A. Ivanovac, V. K. Polovniakd, V. A. Shustovc and Y. V. Yablokov, Delocalization of the Cu^{2+} Unpaired Electron on the Next Nearest Ligands in $Sr_2Pd_{0.99}Cu_{0.01}O_3$ Ceramics, *Acta Physica Polonica*, A, 114 1, 2008, p. 197.

[40]. J. J. Zuckerman (Founding Editor), A. P. Hagen (Editor), Handbook of Inorganic Reactions and Methods, Formation of Bonds to O, S, Se, Te, Po, Part 2, *John Wiley & Sons*, November 1998, 536 pages.

[41]. T. Yamamoto, Synthesis, Structure, and Physical Properties of Novel Iron Oxides Prepared by Topotactic Reactions, Doctoral degree report No. 17235, *Kyoto University*, 2012.

[42]. E. G. Tulsky and J. R. Long, Dimensional reduction: A practical formalism for manipulating solid structures, *Chemistry of Materials,* Vol. 13, 4, 2001, pp. 1149-1166.

[43]. Y. Nishihata, J. Mizuki, T. Akao, H. Tanaka, M. Uenishi, M. Kimura, T. Okamoto, N. Hamda, Self-regeneration of a Pd-perovskite catalyst for automotive emissions control, *Nature,* Vol. 418, 2002, pp. 164-167.

[44]. L. Chen, T. Li, S. Cao, S. Yuan, F. Hong, J. Zhang, The role of 4f-electron on spin reorientation transition of $NdFeO_3$: A first principle study, *Journal of Applied Physics,* Vol. 111, 10, 2012, pp. 103905-103910.

[45]. W. Sławiński, R. Przenioslo, I. Sosnowska, M. Brunelli, M. Bieringer, Anomalous thermal expansion in polycrystalline $NdFeO_3$ studied by SR and X-ray diffraction, *Nuclear Instruments and Methods in Physics Research B,* Vol. 254, 2007, pp. 149-152.

[46]. A. Tiwari, Local environment of Fe in $NdFe1-xNixO_3-\delta$ perovskite oxides, *Journal of Alloys and Compounds,* Vol. 274, 1998, pp. 42-46.

[47]. S. Chanda, S. Saha, A. Dutta, T. P. Sinha, Raman spectroscopy and dielectric properties of nanoceramic $NdFeO_3$, *Materials Research Bulletin,* Vol. 48, 2013, pp. 1688-1693.

[48]. W. Zheng, R. Liu, D. Peng and G. Meng, Hydrothermal synthesis of $LaFeO_3$ under carbonate-containing medium, *Materials Letters,* Vol. 43, 1-2, 2000, pp. 19-22.

[49]. S. S. Manoharam and K. C. Patil, Combustion Route to Fine Particle Perovskite Oxides, *Journal of Solid State Chemistry,* Vol. 102, 1, 1993, pp. 267-276.

[50]. H. Cui, M. Zayat and D. Levy, Epoxide assisted sol–gel synthesis of perovskite-type $LaMxFe1-xO_3$ (M = Ni, Co) nanoparticles, *Journal of Non-Crystalline Solids,* Vol. 352, 28-29, 2006, pp. 3035-3040.

[51]. H. N. Pandya, R. G. Kulkarni and P. H. Parsania, Study of cerium orthoferrite prepared by wet chemical method, *Materials Research Bulletin,* Vol. 25, 8, 1990, pp. 1073-1077.

[52]. S. Singh, A. Singh, B. C. Yadav and P. K. Dwivedi, Fabrication of nanobeads structured perovskite type neodymium iron oxide film: Its structural, optical, electrical and LPG sensing investigations, *Sensors and Actuators B,* Vol. 177, 2013, pp. 730-739.

[53]. N. F. Atta, A. Galal, S. M. Ali, The Effect of the Lanthanide Ion-Type in LnFeO3 on the Catalytic Activity for the Hydrogen Evolution in Acidic Medium, *International Journal of Electrochemical Science,* Vol. 9, 2014, pp. 2132-2148.

[54]. P. Ciambelli, S. Cimino, S. De Rossi, L. Lisi, G. Minelli, P. Porta, G. Russo, $AFeO_3$ (A=La, Nd, Sm) and $LaFe1-xMgxO_3$ perovskites as methane combustion and CO oxidation catalysts: structural, redox and catalytic properties, *Applied Catalysis B: Environmental,* Vol. 29, 2001, pp. 239-250.

[55]. C. Tongyun, S. Liming, L. Feng, Z. Weichang, Z. Qianfeng, C. Xiangfeng, $NdFeO_3$ as anode material for S/O_2 solid oxide fuel cells, *Journal of Rare Earths,* Vol. 30, 11, 2012, pp. 1138-1141.

[56]. H. T. Giang, H. T. Duy, P. Q. Ngan, G. H. Thai, D. T. A. Thu, D. T. Thu, N. N. Toan, Hydrocarbon gas sensing of nano-crystalline perovskite oxides $LnFeO_3$ (Ln = La, Nd and Sm), *Sensors and Actuators B,* Vol. 158, 2011, pp. 246-251.

[57]. Z. Ru, H. Jifan, H. Zhouxiang, Z. Ma, W. Zhanlei, Z. Yongjia, Q. Hongwei, Electrical and CO-sensing properties of $NdFe1-xCoxO_3$ perovskite system, *Journal of Rare Earths,* Vol. 28, 4, 2010, pp. 591-595.

[58]. X. Liu, J. Hu, B. Cheng, H. Qin, M. Jiang, Preparation and gas sensing characteristics of p-type semiconducting $LnFe0.9Mg0.1O_3$ (Ln = Nd, Sm, Gd and Dy) materials, *Current Applied Physics,* Vol. 9, 2009, pp. 613-617.

[59]. F. Deganello, G. Marcì, G. Deganello, Citrate–nitrate auto-combustion synthesis of perovskite-type nanopowders: A systematic approach, *Journal of the European Ceramic Society,* Vol. 29, 3, 2009, pp. 439-450.

[60]. Y. Tao, J. Shao, J. Wang, W. G. Wang, Synthesis and properties of $La0.6Sr0.4CoO_3-\delta$ nanopowder, *Journal of Power Sources,* Vol. 85, 2, 2008, pp. 609-614.

[61]. J. H. Park, J. P. Kim, H. T. Kwon, J. Kim, Oxygen permeability, electrical property and stability of $La0. 8Sr0. 2Co0. 2Fe0. 8O_3-\delta$ membrane, *Desalination,* Vol. 233, 1-3, 2008, pp. 73-81.

[62]. K. Rida, A. Benabbas, F. Bouremmad, M. A. Peña, E. Sastre, A. Martínez-Arias, Effect of strontium and cerium doping on the structural characteristics and catalytic activity for C_3H_6 combustion of perovskite $LaCrO_3$ prepared by sol–gel, *Applied Catalysis B: Environmental,* Vol. 84, 3-4, 2008, pp. 457-467.

[63]. R. Hammami, S. Ben Aïssa, H. Batis, Effects of thermal treatment on physico-chemical and catalytic properties of lanthanum manganite $LaMnO_3+y$, *Applied Catalysis A: General,* Vol. 353, 2, 2009, pp. 145-153.

[64]. A. A. Rabelo, M. C. de Macedo, D. M. de Araújo Melo, C. A. Paskocimas, A. E. Martinelli, R. M. do Nascimento, Synthesis and characterization of $La_{1-x}Sr_xMnO_{3\pm\delta}$ powders obtained by the polymeric precursor route, Materials Research, Vol. 14, 1, 2011, pp. 91-96.

[65]. M. Popa, J. M. Calderón-Moreno, Lanthanum cobaltite thin films on stainless steel, *Thin Solid Films,* Vol. 517, 5, 2009, pp. 1530-1533.

[66]. S. Ivanova, A. Senyshyn, E. Zhecheva, K. Tenchev, V. Nickolov, R. Stoyanova, H. Fuess, Effect of the synthesis route on the microstructure and the reducibility of $LaCoO_3$, *Journal of Alloys and Compounds,* Vol. 480, 2, 2009, pp. 279-285.

[67]. J. Wang, A. Manivannan, N. Wu, Sol–gel derived $La0.6Sr0.4CoO_3$ nanoparticles, nanotubes, nanowires and thin films, *Thin Solid Films,* Vol. 517, 2, 2008, pp. 582-587.

[68]. M. Kumar, S. Srikanth, B. Ravikumar, T. C. Alex, S. K. Das, Synthesis of pure and Sr-doped $LaGaO_3$, $LaFeO_3$ and $LaCoO_3$ and Sr, Mg-doped $LaGaO3$ for ITSOFC application using different wet chemical routes, *Materials Chemistry and Physics,* Vol. 113, 2-3, 2009, pp. 803-815.

[69]. M. E. Rivas, J. L. G. Fierro, R. Guil-López, M. A. Peña, V. La Parola, M. R. Goldwasser, Preparation and characterization of nickel-based

mixed-oxides and their performance for catalytic methane decomposition, *Catalysis Today,* Vol. 133-135, 2008, pp. 367-373.

[70]. F. Boschini, A. Rulmont, R. Cloots, B. Vertruyen, Rapid synthesis of submicron crystalline barium zirconate $BaZrO_3$ by precipitation in aqueous basic solution below 100 °C, *Journal of the European Ceramic Society,* Vol. 29, 8, 2008, 1457-1462.

[71]. K. R. Barnard, K. Foger, T. W. Turney, R. D. Williams, Lanthanum cobalt oxide oxidation catalysts derived from mixed hydroxide precursors, *Journal of Catalysis,* Vol. 125, 2, 1990, pp. 265-275.

[72]. R. Pei, X. Chen, Y. Suo, T. Xiao, Q. Ge, H. Yao, J. Wang, Z. Li, Synthesis of $La_{0.85}Sr_{0.15}Ga_{0.8}Mg_{0.2}O_{3-\delta}$ powder by carbonate co-precipitation combining with azeotropic-distillation process, *Solid State Ionics,* Vol. 219, 2012, pp. 34-44.

[73]. N. S. Chae, K. S. Park, Y. S. Yoon, I. S. Yoo, J. S. Kim, H. H. Yoon, Sr- and Mg-doped $LaGaO_3$ powder synthesis by carbonate coprecipitation, *Colloids and Surfaces A: Physicochemical and Engineering Aspects,* Vol. 313-314, 2008, pp. 154-156.

[74]. Q. Jin, Microwave chemistry, *China Science Press,* Beijing, 1999.

[75]. M. P. Selvam, K. Rao, Microwave Synthesis and Consolidation of Gadolinium Aluminum Perovskite, a Ceramic Extraordinaire, *Advanced Materials,* Vol. 12, 2000, pp. 1621-1624.

[76]. K. E. Gibbons, S. J. Blundell, A. I. Mihut, I. Gameson, P. P. Edwards, Y. Miyazaki, N. C. Hyatt, M. O. Jones, A. Porch, Hydrolysis, oxidation and complexation: the reactions of tin(II) chloride with a tripodal Schiff base ligand, *Chemical Communications,* Vol. 1, 2000, pp. 159-160.

[77]. M. P. Selvam, K. J. Rao, Microwave preparation and sintering of industrially important perovskite oxides: $LaMO_3$ (M = Cr, Co, Ni), *Journal of Materials and Chemistry,* Vol. 13, 2003, pp. 596-601.

[78]. H. Yan, X. Huang, Z. Lu, H. Hu, R. Xue, L. Chen, Microwave synthesis of $LiCoO_2$ cathode materials, *Journal of Power Sources,* Vol. 68, 1997, pp. 530-532.

[79]. L. S. Cavalcante, V. S. Marques, J. C. Sczancoski, M. T. Escote, M. R. Joya, J. A. Varela, M. R. M. C. Santos, P. S. Pizani, E. Longo, Synthesis, structural refinement and optical behavior of $CaTiO_3$ powders: A comparative study of processing in different furnaces, *Chemical Engineering Journal,* Vol. 143, 1-3, 2008, pp. 299-307.

[80]. A. Kaddouri, P. Gelin, N. Dupont, Methane catalytic combustion over La–Ce–Mn–O- perovskite prepared using dielectric heating, *Catalysis Communications,* Vol. 10, 7, 2009, pp. 1085-1089.

[81]. S. Farhadi, Z. Momeni, M. Taherimehr, Rapid synthesis of perovskite-type $LaFeO_3$ nanoparticles by microwave-assisted decomposition of bimetallic $La[Fe(CN)_6]\cdot5H_2O$ compound, *Journal of Alloys and Compounds,* Vol. 471, 1-2, 2009, pp. L5-L8.

[82]. N. Charoenthai, R. Traiphol, G. Rujijanagul, Microwave synthesis of barium iron niobate and dielectric properties, *Materials Letters,* Vol. 62, 29, 2008, pp. 4446-4448.

[83]. A. J. Paula, R. Parra, M. A. Zaghete, J. A. Varela, Synthesis of KNbO₃ nanostructures by a microwave assisted hydrothermal method, *Materials Letters*, Vol. 62, 17-18, 2008, pp. 2581-2584.

[84]. J. H. Ryu, S. Koo, D. S. Chang, J. Yoon, C. S. Lim, K. B. Shim, Microwave-assisted synthesis of PbWO₄ nano-powders via a citrate complex precursor and its photoluminescence, *Ceramics International*, Vol. 32, 6, 2006, pp. 647-652.

[85]. J. H. Ryu, S. Koo, D. S. Chang, J. Yoon, C. S. Lim, K. B. Shim, Microwave-assisted synthesis of CaMoO₄ nano-powders by a citrate complex method and its photoluminescence property, *Journal of Alloys and Compounds*, Vol. 390, 1-2, 2005, pp. 245-249.

[86]. J. H. Ryu, J. Yoon, C. S. Lim, W. Oh, K. B. Shim, Microwave-assisted synthesis of nanocrystalline MWO₄ (M: Ca, Ni) via water-based citrate complex precursor, *Ceramics International*, Vol. 31, 6, 2005, pp. 883-888.

[87]. Y. Fu, C. Lin, Fe/Sr ratio effect on magnetic properties of strontium ferrite powders synthesized by microwave-induced combustion process, *Journal of Alloys and Compounds*, Vol. 386, 1-2, 2005, pp. 222-227.

[88]. Y. M. Abd Al-Rahman. Characterization and some Applications of Nano-Inorganic Oxides Synthesized by Microwave Technique, M. Sc. Thesis, Faculty of Science, *Cairo University*, 2013.

[89]. P. V. Gosavi, R. B. Biniwale, Pure phase LaFeO₃ perovskite with improved surface area synthesized using different routes and its characterization, *Materials Chemistry and Physics*, Vol. 119, 2010, pp. 324-329.

[90]. C. Vijayakumar, H. P. Kumar, S. Solomon, J. K. Thomas, P. R. S. Warrior, J. Koshy, Synthesis, characterization, sintering and dielectric properties of nanostructured perovskite-type oxide, Ba₂GdSbO₆, *Bulletin Materials Science*, Vol. 31, 5, 2008, pp. 719-722.

[91]. N. F. Atta, S. M. Ali, E. H. El-Ads, A. Galal, Nano-perovskite carbon paste composite electrode for the simultaneous determination of dopamine, ascorbic acid and uric acid, *Electrochimica Acta*, Vol. 128, 2014, pp. 16-24.

[92]. N. F. Atta, E. H. El-Ads, A. Galal, Evidence of Core-Shell Formation between NdFeO₃ Nano-Perovskite and Ionic Liquid Crystal and Its Application in Electrochemical Sensing of Metoclopramide, *Journal of The Electrochemical Society*, Vol. 163 , 7, 2016, pp. B325-B334.

[93]. A. Galal, S. A. Darwish, N. F. Atta, S. M. Ali, A. A. Abd El Fatah, Synthesis, structure and catalytic activity of nano-structured Sr–Ru–O type perovskite for hydrogen production, *Applied Catalysis A: General*, Vol. 378, 2010, pp. 151-159.

[94]. A. Galal, N. F. Atta, S. M. Ali, Investigation of the catalytic activity of LaBO₃ (B = Ni, Co, Fe or Mn) prepared by the microwave-assisted method for hydrogen evolution in acidic medium, *Electrochimica Acta*, Vol. 56, 2011, pp. 5722-5730.

[95]. A. Galal, N. F. Atta, S. M. Ali, Optimization of the synthesis conditions for LaNiO₃ catalyst by microwave assisted citrate method for hydrogen

production, *Applied Catalysis A: General*, Vol. 409–410, 2011, pp. 202-208.

[96]. N. F. Atta, A. Galal, S. M. Ali, The Catalytic Activity of Ruthenates ARuO$_3$ (A= Ca, Sr or Ba) for the Hydrogen Evolution Reaction in Acidic Medium, *International Journal of Electrochemical Science*, Vol. 7, 2012, pp. 725-746.

[97]. S. M. Ali, Y. M. Abd Al-Rahman, A. Galal, Catalytic Activity toward Oxygen Evolution of LaFeO$_3$ Prepared by the Microwave Assisted Citrate Method, *Journal of Electrochemical Society*, Vol. 159, 9, 2012, pp. F600-F605.

[98]. A. Galal, N. F. Atta, S. A. Darwish, A. A. Abd El Fatah, S. M. Ali, Electrocatalytic evolution of hydrogen on a novel SrPdO$_3$ perovskite electrode, *Journal of Power Sources*, Vol. 195, 2010, pp. 3806-3809.

[99]. S. Ghosh, S. Dasgupta, Synthesis, characterization and properties of nanocrystalline perovskite cathode materials, *Materials Science-Poland*, Vol. 28, 2, 2010, pp. 427-438.

[100]. G. Viruthagiri, P. Praveen, S. Mugundan, E. Gopinathan, Synthesis and Characterization of Pure and Nickel doped SrTiO$_3$ Nanoparticles via Solid State Reaction Route, *Indian Journal of Advances in Chemical Science*, Vol. 1, 3, 2013, pp. 132-138.

[101]. S. Li. Preparation and characterization of perovskite structure lanthanum gallate and lanthanum aluminate based oxides, PhD Thesis, *Royal Institute of Technology*, Stockholm, Sweden, 2009.

[102]. G. Pecchi, C. M. Campos, M. G. Jiliberto, E. J. Delgado, J. L. G. Fierro, Effect of additive Ag on the physicochemical and catalytic properties of LaMn0.9Co0.1O3.5 perovskite, *Applied Catalysis A: General*, Vol. 371, 2009, pp. 78-84.

[103]. M. Khorasani-Motlagh, M. Noroozifar, M. Yousefi, S. Jahani, Chemical Synthesis and Characterization of Perovskite NdfeO$_3$ Nanocrystals via a Co-Precipitation Method, *International Journal of Nanoscience and Nanotechnology*, Vol. 9, 1, 2013, pp. 7-14.

[104]. M. Carolina Navarro, Elisa V. Pannunzio-Miner, Silvina Pagola, M. Inés Gómez, Raúl E. Carbonio, Structural refinement of Nd[Fe(CN)$_6$]·4H$_2$O and study of NdFeO3 obtained by its oxidative thermal decomposition at very low temperatures, *Journal of Solid State Chemistry*, Vol. 178, 2005, pp. 847-854.

[105]. N. Lakshminarayanan, H. Choi, J. N. Kuhn, U. S. Ozkan, Effect of additional B-site transition metal doping on oxygen transport and activation characteristics in La0.6Sr0.4(Co0.18Fe0.72X0.1)O3−δ (where X = Zn, Ni or Cu) perovskite oxides, *Applied Catalysis B: Environmental*, Vol. 103, 2011, pp. 318-325.

[106]. Y. Cho, K. Choi, Y. Kim, J. Jung, S. Lee, Characterization and Catalytic Properties of Surface La-rich LaFeO$_3$ Perovskite, *Bulletin of the Korean Chemical Society*, Vol. 30, 6, 2009, pp. 1368-1372.

[107]. D. A. Pawlak, M. Ito, M. Oku, K. Shimamura, T. Fukuda, Interpretation of XPS O (1s) in Mixed Oxides Proved on Mixed Perovskite Crystal, *Journal of Physical Chemistry B*, Vol. 106, 2002, pp. 504-507.

[108]. A. Ito, H. Masumoto, T. Goto, Microstructure and Electrical Conductivity of $SrRuO_3$ Thin Films Prepared by Laser Ablation, *Materials Transactions*, Vol. 47, 11, 2006, pp. 2808-2814.

[109]. C. M. Pradier, C. Hinnen, K. Jansson, L. Dahl, M. Nygren, A. Flodstrom, Structural and surface characterization of perovskite-type oxides, influence of A and B substitutions upon oxygen binding energy, *Journal of Materials Science*, Vol. 33, 1998, pp. 3187-3191.

[110]. Q. Wu, M. Liu, W. Jaegermann, X-ray photoelectron spectroscopy of $La_{0.5}Sr_{0.5}MnO_3$, *Materials Letters*, Vol. 59, 16, 2005, pp. 1980-1983.

[111]. C. Li, C. Wang and Y. Lin, Pd-integrated lanthanum-transition metal perovskites for methanol partial oxidation, *Catalysis Today*, Vol. 174, 2011, pp. 135-140.

[112]. M. Ghasdi, H. Alamdari, CO sensitive nanocrystalline $LaCoO_3$ perovskite sensor prepared by high energy ball milling, *Sensors and Actuators B*, Vol. 148, 2010, pp. 478-485.

[113]. M. Ghasdi, H. Alamdari, S. Royer, A. Adnot, Electrical and CO gas sensing properties of nanostructured $La_{1-x}Ce_xCoO_3$ perovskite prepared by activated reactive synthesis, *Sensors and Actuators B*, Vol. 156, 2011, pp. 147-155.

[114]. Z. Ru, H. Jifan, H. Zhouxiang, Z. Ma, W. Zhanlei, Z. Yongjia, Q. Hongwei, Electrical and CO-sensing properties of $NdFe_{1-x}Co_xO_3$ perovskite system, *Journal of Rare Earths*, Vol. 28, 4, 2010, pp. 591-595.

[115]. J. W. Fergus, Perovskite oxides for semiconductor-based gas sensors, *Sensors and Actuators B*, Vol. 123, 2007, pp. 1169-1179.

[116]. P. Song, Q. Wang, Z. Yang, The effects of annealing temperature on the CO-sensing property of perovskite $La_{0.8}Pb_{0.2}Fe_{0.8}Cu_{0.2}O_3$ nanoparticles, *Sensors and Actuators B*, Vol. 141, 2009, pp. 109-115.

[117]. C. Doroftei, P. D. Pop, F. Iacomi, Synthesis of nanocrystalline La–Pb–Fe–O perovskite and methanol-sensing characteristics, *Sensors and Actuators B*, Vol. 161, 1, 2011, pp. 977-981.

[118]. A. Benali, S. Azizi, M. Bejar, E. Dhahri, M. F. P. Graça, Structural, electrical and ethanol sensing properties of double-doping $LaFeO_3$ perovskite oxides, *Ceramics International*, Vol. 40, 9, 2014, pp. 14367-14373.

[119]. X. Liu, B. Cheng, J. Hu, H. Qin, M. Jiang, Semiconducting gas sensor for ethanol based on $LaMg_xFe_{1-x}O_3$ nanocrystals, *Sensors and Actuators B*, Vol. 129, 2008, pp. 53-58.

[120] Y. Wang, J. Chen, X. Wu, Preparation and gas-sensing properties of perovskite-type $SrFeO_3$ oxide, *Materials Letters*, Vol. 49, 2001, pp. 361-364.

[121]. X. Liu, J. Hu, B. Cheng, H. Qin, M. Jiang, Preparation and gas sensing characteristics of p-type semiconducting $LnFe_{0.9}Mg_{0.1}O_3$ (Ln = Nd, Sm,

Gd and Dy) materials, *Current Applied Physics*, Vol. 9, 2009, pp. 613-617.

[122]. H. Najjar, H. Batis, La–Mn perovskite-type oxide prepared by combustion method: Catalytic activity in ethanol oxidation, *Applied Catalysis A: General*, Vol. 383, 2010, pp. 192-201.

[123]. L. Sun, H. W. Qin, E. Cao, M. Zhao, F. Gao, J. Hu, Gas-sensing properties of perovskite $La_{0.875}Ba_{0.125}FeO_3$ nanocrystalline powders, *Journal of Physics and Chemistry of Solids*, Vol. 72, 2011, pp. 29-33.

[124]. L. Kong, Y. Shen, Gas-sensing property and mechanism of $Ca_xLa_{1-x}FeO_3$ ceramics, *Sensors and Actuators B*, Vol. 30, 1996, pp. 217-221.

[125]. C. Feng, S. Ruan, J. Li, B. Zou, J. Luo, W. Chen, W. Dong, F. Wu, Ethanol sensing properties of $LaCo_xFe_{1-x}O_3$ nanoparticles: Effects of calcination temperature, Co-doping, and carbon nanotube-treatment, *Sensors and Actuators B*, Vol. 155, 2011, pp. 232-238.

[126]. S. M. Khetre, Ethanol Gas Sensing Properties of Nano-Porous $LaFeO_3$ Thick Film, *Sensors and Transducers*, Vol. 149, 2, 2013, pp. 13-19.

[127]. L. Zhang, J. Hu, P. Song, H. Qin, M. Jiang, Electrical properties and ethanol-sensing characteristics of perovskite $La_{1-x}Pb_xFeO_3$, *Sensors and Actuators B*, Vol. 114, 2006, pp. 836-840.

[128]. G. Martinelli, M. C. Carotta, M. Ferroni, Y. Sadaoka, E. Traversa, Screen-printed perovskite-type thick films as gas sensors for environmental monitoring, *Sensors and Actuators B*, Vol. 55, 1999, pp. 99-110.

[129]. M. C. Carotta, M. A. Butturi, G. Martinelli, Y. Sadaoka, P. Nunziante, E. Traversa, Microstructural evolution of nanosized $LaFeO_3$ powders from the thermal decomposition of a cyano-complex for thick film gas sensors, *Sensors and Actuators B*, Vol. 44, 1997, pp. 590-594.

[130]. E. Di Bartolomeo, M. L. Grilli, J. W. Yoon, E. Traversa, NO_x sensors based on interfacing nano-sized $LaFeO_3$ perovskite-type oxide and ionic conductors. Project of Special Materials for Advanced Technologies, *J. Electrochem. Soc.*, Vol. 148, Issue 9, 2001, pp. H98-H102.

[131]. K. Sahner, R. Moos, M. Matam, J. J. Tunney, M. Post, Hydrocarbon sensing with thick and thin film p-type conducting perovskite materials, *Sensors and Actuators B*, Vol. 108, 2005, pp. 102-112.

[132]. H. T. Giang, H. T. Duy, P. Q. Ngan, G. H. Thai, D. T. A. Thu, D. T. Thu, N. N. Toan, Hydrocarbon gas sensing of nano-crystalline perovskite oxides $LnFeO_3$ (*Ln* = La, Nd and Sm), *Sensors and Actuators B*, Vol. 158, 2011, pp. 246-251.

[133]. T. Addabbo, F. Bertocci, A. Fort, M. Gregorkiewitz, M. Mugnaini, R. Spinicci, V. Vignoli, Gas sensing properties and modeling of $YCoO_3$ based perovskite materials, *Sensors and Actuators B*, Vol. 221, 2015, pp. 1137-1155.

[134]. C. Balamurugan, S. Arunkumar, D. W. Lee, Hierarchical 3D nanostructure of $GdInO_3$ and reduced-graphene-decorated $GdInO_3$

nanocomposite for CO sensing applications, *Sensors and Actuators B*, Vol. 234, 2016, pp. 155-166.

[135]. A. Kumar, Y. Rajneesh, K. Singh, P. Singh, Fabrication of lanthanum ferrite based liquefied petroleum gas sensor, *Sensors and Actuators B*, Vol. 229, 2016, pp. 25-30.

[136]. F. Jia, H. Zhong, W. Zhang, X. Li, G. Wanga, J. Songa, Z. Cheng, J. Yin, L. Guo, A novel nonenzymatic ECL glucose sensor based on perovskite LaTiO$_3$-Ag0.1 nanomaterials, *Sensors and Actuators B*, Vol. 212, 2015, pp. 174-182.

[137]. D. Ye, Y. Xu, L. Luo, Y. Ding, Y. Wang, X. Liu, L. Xing, J. Peng, A novel nonenzymatic hydrogen peroxide sensor based on LaNi$_{0.5}$Ti$_{0.5}$O$_3$/CoFe$_2$O$_4$ modified electrode, *Colloids and Surfaces B: Biointerfaces*, Vol. 89, 2012, pp. 10-14.

[138]. Z. Zhang, S. Gu, Y. Ding, J. Jin, A novel nonenzymatic sensor based on LaNi$_{0.6}$Co$_{0.4}$O$_3$ modified electrode for hydrogen peroxide and glucose, *Analytica Chimica Acta*, Vol. 745, 2012, pp. 112-117.

[139]. G. L. Luque, N. F. Ferreyra, A. G. Leyva, G. A. Rivas, Characterization of carbon paste electrodes modified with manganese based perovskites-type oxides from the amperometric determination of hydrogen peroxide, *Sensors and Actuators B*, Vol. 142, 2009, pp. 331-336.

[140]. Z. Zhang, S. Gu, Y. Ding, F. Zhang, J. Jin, Determination of hydrogen peroxide and glucose using a novel sensor platform based on Co$_{0.4}$Fe$_{0.6}$LaO$_3$ nanoparticles, *Microchimica Acta*, Vol. 180, 2013, pp. 1043-1049.

[141]. Y. Wang, H. Zhong, X. Li, F. Jia, Y. Shi, W. Zhang, Z. Cheng, L. Zhang, J. Wang, Perovskite LaTiO$_3$–Ag0. 2 nanomaterials for nonenzymatic glucose sensor with high performance, *Biosensors and Bioelectronics*, Vol. 48, 2013, pp. 56-60.

[142]. Y. Wang, Y. Xu, L. Luo, Y. Ding, X. Liu, Preparation of perovskite-type composite oxide LaNi$_{0.5}$Ti$_{0.5}$O$_3$–NiFe$_2$O$_4$ and its application in glucose biosensor, *Journal of Electroanalytical Chemistry*, Vol. 642, 2010, pp. 35-40.

[143]. Y. Shimizu, H. Komatsu, S. Michishita, N. Miura, N. Yamazo, Sensing characteristics of hydrogen peroxide sensor using carbon-based electrode loaded with perovskite-type oxide, *Sensors and Actuators B*, Vol. 34, 1996, pp. 493-498.

[144]. D. T. V. Anh, W. Olthuis, P. Bergveld, Sensing properties of perovskite oxide La$_{1-x}$Sr$_x$CoO$_{3-\delta}$ obtained by using pulsed laser deposition, *Sensors and Actuators B: Chemical*, 103, 1-2, 2004 pp. 165-168.

[145]. E. H. El-Ads, A. Galal, N. F. Atta, Electrochemistry of glucose at gold nanoparticles modified graphite/SrPdO$_3$ electrode – Towards a novel non-enzymatic glucose sensor, *Journal of Electroanalytical Chemistry*, Vol. 749, 2015, pp. 42-52.

[146]. C. Y. Wen, Study and Application of Composite Materials Based on Perovskite Nanoparticles, M. Sc. Thesis, *Jiangsu University of Science and Technology*, 2011.

[147]. B. Wang, S. Gu, Y. Ding, Y. Chu, Z. Zhang, X. Ba, Q. Zhang, X. Li, A novel route to prepare $LaNiO_3$ perovskite-type oxide nanofibers by electrospinning for glucose and hydrogen peroxide sensing, *Analyst*, Vol. 138, 1, 2013, pp. 362-367.

[148]. D. Xu, L. Luo, Y. Ding, P. Xu, Sensitive electrochemical detection of glucose based on electrospun $La_{0.88}Sr_{0.12}MnO_3$ naonofibers modified electrode, *Analytical Biochemistry*, Vol. 489, 2015, pp. 38-43.

[149]. B. Cai, M. Zhao, Y. Wang, Y. Zhou, H. Cai, Z. Ye, J. Huang, A perovskite-type $KNbO_3$ nanoneedles based biosensor for direct electrochemistry of hydrogen peroxide, *Ceramics International*, Vol. 140, 2014, pp. 8111-8116.

[150]. F. Deganello, L. F. Liotta, S. G. Leonardi, G. Neri, Kinetics of Pb and Pb-Ag anodes for zinc electrowinning—I. Formation of $PbSO_4$ layers at low polarization, *Electrochimica Acta*, Vol. 190, 2015, pp. 939-948.

[151]. S. Thirumalairajan, K. Girija, V. R. Mastelaro, V. Ganesh, N. Ponpandian, Detection of neurotransmitter compound dopamine by modified glassy carbon electrode with self-assembled perovskite $LaFeO_3$ microspheres constructed of nanospheres, *RSC Advances*, Vol. 4, 2014, pp. 25957- 25962.

[152]. N. F. Atta, S. M. Ali, E. H. El-Ads, A. Galal, The Electrochemistry and Determination of Some Neurotransmitters at $SrPdO_3$ Modified Graphite Electrode, *Journal of the Electrochemical Society*, Vol. 160, 7, 2013, pp. G3144-G3151.

[153]. S. Thirumalairajan, K. Girija, V. Ganesh, D. Mangalaraj, C. Viswanathan, N. Ponpandian, Novel Synthesis of $LaFeO_3$ Nanostructure Dendrites: A Systematic Investigation of Growth Mechanism, Properties, and Biosensing for Highly Selective Determination of Neurotransmitter Compounds, *Crystal Growth and Design*, Vol. 13, 2013, pp. 291-302.

[154]. G. Wang, J. Sun, W. Zhang, S. Jiao, B. Fang, Simultaneous determination of dopamine, uric acid and ascorbic acid with $LaFeO_3$ nanoparticles modified electrode, *Microchimica Acta*, Vol. 164, 2009, pp. 357-362.

[155]. S. Priyatharshni, M. Divagar, C. Viswanathan, D. Mangalaraj, N. Ponpandian, Electrochemical Simultaneous Detection of Dopamine, Ascorbic Acid and Uric Acid Using $LaMnO_3$ Nanostructures, *Journal of The Electrochemical Society*, Vol. 163, 8, 2016, pp. B460-B465.

[156]. T. R. L. Dadamos, B. H. Freitas, D. H. M. Genova, R. D. Espirito-Santo, E. R. P. Gonzalez, S. Lanfredi, M. F. S. Teixeira, Electrochemical characterization of the paste carbon modified electrode with $KSr_2Ni_{0.75}Nb_{4.25}O_{15-\delta}$ solid in catalytic oxidation of the dipyrone, *Sensors and Actuators B*, Vol. 169, 2012, pp. 267-273.

[157]. Y. Wang, L. Luo, Y. Ding, X. Zhang, Y. Xu, X. Liu, Perovskite nanoparticles $LaNi_{0.5}Ti_{0.5}O_3$ modified carbon paste electrode as a sensor for electrooxidation and detection of amino acids, *Journal of Electroanalytical Chemistry*, Vol. 667, 2012, pp. 54-58.

[158]. L. Bian, J. Xu, M. Song, F. Dong, H. Dong, F. Shi, L. Wang, W. Ren, Designing perovskite BFO (111) membrane as an electrochemical sensor for detection of amino acids: A simulation study, *Journal of Molecular Structure*, Vol. 1099, 2015, pp. 1-9.

5.

Functionalized Carbon Based Materials for Sensing and Biosensing Applications: from Graphite to Graphene

Nada F. Atta, Hagar K. Hassan and Ahmed Galal

5.1. Introduction

Carbon, the element of life, due to its wide abundance and using, carbon can be considered as one of the most attractive element; you can find it wherever you go, in air, in wood, in meals, even on your furniture and clothes. It is also found in earth's crust, sun, stars, comets and the atmosphere of planets and it represents the forth abundant element in the space after hydrogen, helium and oxygen [1]. So, it deserves to be the element of life worthily. The scientists knew this fact well so, for many years the carbon allotropes have been used in a wide range of applications including energy conversion, energy storage, environmental aspects, field emission displays, electronics and sensors [2]. Carbon differs than any other element in the periodic table in its large number of structures and allotropes depending on the way by which its atoms are arranged. Those carbon allotropes are completely different in their chemical and physical properties. The simplest example for carbon allotropes divergence is clearly shown in graphite and diamond. As well known, graphite is the softness material while diamond is the hardest one [1].

Construction of new sensors based on the physical or chemical modifications of electrodes has attracted the attention of scientists' long time ago [2]. Functionalization of the electrode materials including carbonaceous materials is very crucial. One of the aims of the electrode functionalization is to enhance the electrocatalytic activity toward the analyte through promoting the charge transfer process and at the same time to reduce the reactivity toward some other species that are known by "interfering materials" [3]. Additionally, the electrode functionalization may be used to increase the resistivity of the electrode surface toward surface fouling resulting in enhancement of the long-term stability of the electrode. So, we can generally say that; functionalization of electrodes increases the reactivity, stability, and selectivity of the electrode material.

One of the disadvantages of the carbon materials is its poor solubility that can be tuned by surface functionalization [4]. There are many approaches for functionalization of carbon materials but we can generally divide them into two main routes: according to the reaction medium into wet or dry methods or according to the nature of bond formed between carbon and modifier into a covalent or non-covalent method. Covalent functionalization is based on the formation of a covalent bond between carbon and other components. Non-covalent functionalization is mainly based on the adsorption via van der Waals force, hydrogen bonds, electrostatic force and π-stacking interactions [5]. Some of the functional entities that are usually used for carbon functionalization are the polymeric materials, biomolecules, nanoparticles, element doping, or by introducing an organic group to their surfaces which is usually oxygen-containing one [6]. The scope of this chapter is to highlight in details the methods of carbon functionalization and their applications in sensors and biosensors.

5.2. Graphite and Carbon Paste

The word graphite is driven from the Greek word "graphein" which means "to write". This name is given to graphite because it has been used in writing. Although graphite is known since the early centuries, it is only demonstrated as an allotrope of carbon in the 18th century [7].

Graphite structure is considered as stacked layers of graphene sheets connected together via van der Waal forces and some π-bonds as confirmed by SPM imaging [8]. XRD of graphite shows that there are two graphitic structures. The hexagonal is the most dominant and stable graphite form with ABAB stacking. And the rhombohedral structural that is thermodynamically unstable and present in small content in combination with the hexagonal form and its amount can be reduced by the high-temperature treatment of graphite above 1300 °C [8, 9]. The density of both structures is 2.26 g/cm^3. C-C bond distance in graphite is 142 pm which is intermediate between C-C single bond and C=C lengths while the interlayer distance in graphite is 335 pm. This large distance confirms the assumption of connecting the graphitic layer together by van der Waals forces [9]. Actually, the ideal graphitic structure doesn't exist but rather there are many graphitic materials that are considered as aggregates of crystallites of graphite. Those graphitic materials such as pyrolytic graphite, carbon black, vitreous carbon, etc. differ in their sizes, imperfections and considerably in their properties [8]. On the next sections we are going to talking about these graphitic structures and their functionalization or modifications and their use in sensing applications.

Carbon paste (CP) is considered as a modification of graphite powder with paraffin oil to get the paste shape where graphite powder is mixed with paraffin oil using a mortar and pestle and then the resulted paste is packed into the electrode tube. At the end of 1950, carbon paste electrode was firstly introduced by Adams [10]. However, its functionalization was started few years later by Kuwana and French [11]. The functionalization or modification of CPE is usually performed by two routs; either by functionalization of its surface or by incorporation of the modifier into the interior of CP through a physical mixing process as well be discussed in the next sections.

5.2.1. Polymeric Functionalization

Functionalization of the graphite or carbon paste (CP) electrodes with a polymeric material can be achieved by different routes; electro-polymerization of the polymer on the surface of the electrode, in situ chemical polymerization in the presence of a substrate or by the outside chemical or thermal polymerization followed by mixing with CP. Many polymeric materials have been used for modification of carbon materials.

Electropolymerization is one of the most known methods for polymeric functionalization of any electrode surface. Electropolymerization process under a controlled potential can be performed by either applying a constant potential (under which the polymerization occurs) or by cycling voltammetry technique (CV), galvanostatic method is also possible. Determination of the potential under which the polymerization takes place and its optimization is so crucial whatever a constant potential technique is used or CV. So, it is important to perform a potential scan to know exactly the polymerization potential on the selected material or substrate. One of the most predominant methods for the carbon paste functionalization is by attaching a polymeric chain to its surface. One of the works concerning the electro-polymerization of a monomer on the surface of CP is that published in 1996 by E. Miland et al. [12]. In this work, O-aminophenol was electro-polymerized on the surface of well-polished CPE mixed with 7 % (w/w) uricase and 3 % (w/w) Horseradish peroxidase (HRP) and used as an enzymatic sensor for uric acid determination. The electropolymerization of O-aminophenol was performed using CV technique in 0.1 M acetic acid (pH 5) from 0.0 to 0.7 V vs Ag/AgCl reference electrode and using scan rate of 50 mV s^{-1}. The electrolyte was subjected to gas purging before the polymerization step to remove the molecular oxygen. They found that this bienzymatic modified poly (o-aminophenol) CPE showed a linear response up to 1×10^{-4} M with a detection limit of 3×10^{-6} M. The role of poly (o-aminophenol) as they stated is to increase the resistance of the electrode toward the interfering species as well as the surface fouling.

Gupta and Goyal designed a sensor for serotonin based on the modified pyrolytic graphite with polymelamine [14]. Their sensor is based on the electropolymerization of melamine on the surface of pyrolytic graphite. Its polymerization was performed by the electropolymerization technique in an acidic electrolyte (1.0 mM melamine in 0.1 M H$_2$SO$_4$) in a potential range from 0.1 to 1.6 V (vs. Ag/AgCl reference electrode), scan rate 100 mV s^{-1} for 20 cycles. The limit of detection of serotonin on polymelamine/pyrolytic graphite electrode is 30 nM in urine and serum with a sensitivity of 0.088 µA µM^{-1}.

Moreover, one of the most selective sensors is that based on the molecular imprinting technique, in which a selected monomer was polymerized in the presence of a certain analyte (template) to form polymer-template structure. This process is followed by an extraction process leaving the polymer with a cavity which is suitable only for the analyte molecule. This is similar to the idea of lock and key as shown in Fig. 5.1.

Nezhadali et al [15] used the molecular imprinting polymer (MIP) to modify graphite pencil electrode for benzimidazole determination. Their method is based on the electropolymerization of pyrrole on 0.7 mm graphite pencil in the presence of benzimidazole followed by removing benzimidazole from the prepared polymer leaving polymer with a cavities suitable only for benzimidazole. The electropolymerization process was performed using CV technique from -1.0 to 1.9 V vs. Ag/AgCl, scan rate of 100 mV s^{-1} for 25 cycles in a polymerization bath containing 0.1 M NaClO$_4$, 0.02 M pyrrole, and 2.5 mM benzimidazole.

The embedded benziimdazole was removed from the polymer film through performing DPV from 0.1 to -0.9 V in Britton-Robinson buffer solution of pH 5.0. They also used biological model samples of beef, turkey, chicken, lamb and fish meats. The used sensor recorded a detection limit of 7.0 × 10^{-7} over the linear range from 3×10^{-6} to 5×10^{-3} M.

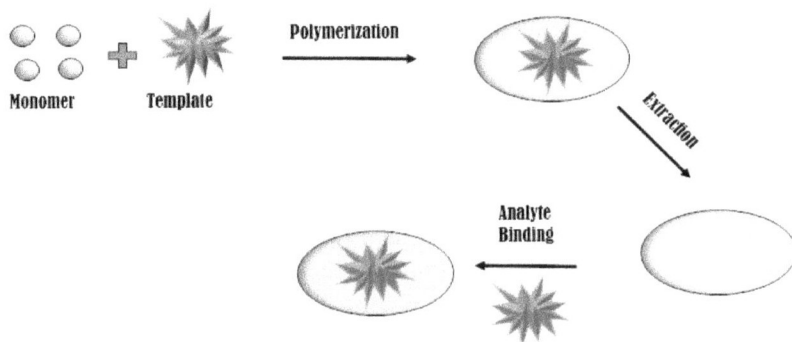

Fig. 5.1. Schematic representations of molecular imprinting technique showing inclusion and extraction of template molecule.

Fractionalization of CPE by incorporation of a MIP for the selective determination of propylparaben (PP) was also achieved by Gholivand et al. [16]. In their work, they used PP as a template and methylacrylic acid as a monomer to prepare a MIP/CPE sensor for PP determination in cosmetics. Parabens compounds are restricted to be used in many cosmetic products with a percentage higher than 0.4 % (w/w) for a single ester and 0.8 % (w/w) for ester mixture so it is very important to design a sensor that could be used for the determination of its concentration in cosmetics as low as possible. MIP/CPE achieved a very high selectivity toward PP and a very low detection limit as low as 0.32 nM in the linear range from 1 nM to 100 nM. Thermal polymerization technique was used for MIP preparation. Briefly, 0.1 mmol PP, 0.4 mmol methacrylic acid, 10 ml acetonitrile in dimethylsulfoxide (66.5 % v/v) as a polymerization solvent, 2 mmol EGDMA as a cross-linker and finally 0.1 mmol AIBN as an initiator are mixed together in a screw-capped glass vial then it is sealed under nitrogen and was left in a water bath at 60 °C for 24 hours with an occasional stirring. The removal of PP from the polymer was successfully obtained by cleaning it with methanol/acetic acid mixture.

Another work that used MIP/CPE as a sensor is that achieved by Sadeghi et al [17] for determination of sulfadiazine in cow milk and human serum samples. In their work, they used sulfadiazine as a template and its extraction from the polymeric material was achieved by cleaning it with methanol/acetic acid solution mixture (90:10 % v/v). The optimum percentage of MIP was 18 %wt, 57 %wt graphite powder and 25 %wt n-eicosane (binder). The calculated LOD and the sensitivity are 1.4×10^{-7} M and 4.2×10^{5} μA M^{-1}, respectively. The advantages of this sensor in addition to its high sensitivity, selectivity and the low detection limit are its high stability where there were no significant changes in the performance after 18 weeks. Benzo[a]pyrene (BaP) does not have any functional groups and has a very high oxidation potential (0.94 V vs calomel electrode) [18]. So, its direct electrochemical oxidation in aqueous media is restricted and finding another method for its determination has a significant importance. Recently, Udomsap et al [18] introduced a method for the determination of benzo[a]pyrene based on the electrochemical molecular imprinted polymer (eMIP/CPE). Their method is based on using redox tracers (vinyl ferrocene) in order to promote the recognition of BaP by MIP. Two MIPCPEs were prepared according to the amount of the solvent added (toluene); eMIP20CPE and eMIP30CPE both of them provided a higher performance for BaP determination compared to the non-imprinted polymer (NIP). The detection limits calculated for both NIP and MIP are 0.25 and 0.09 μM, respectively. As the authors mentioned in their study, CP was chosen as an immobilizing material for eMIP because of its high stability in various solvents, low background and residual current and its ease of preparation. More information about MIP and its application in sensing and biosensing are present in [19, 20].

Polysiloxane and mesoporous silica are commonly used to modify CPE for metal sensing application. In 2003, Yantasee et al [21] constructed a uranium sensor based on CPE modified with carbamoylphosphonic acid Functionalized Mesoporous Silica as a new mercury-free metal sensor.

5.2.2. Metal and Metal Oxide Nanoparticles

Gold nanoparticles (AuNPs) are the widely used nanoparticles in sensing applications. Modification of carbon materials with AuNPs attracts a great of attention due to the large active sites and high conductivity of AuNPs [22]. For example, Karuppiah et al. [23] decorated activated graphite with AuNPs for hydrazine determination. AG/Au-Nps/ screen-printed carbon electrode (SPCE) composite modified electrode showed a higher catalytic response and lower overpotential compared to Au-Nps/SPCE and/or graphite /SPCE. The measured limit of detection is 0.57 ± 0.03 nM and a wide linear range from 0.002 µM to 936 µM has been obtained. The AG/SPCE was successfully fabricated by drop casting a 8 µl of graphite suspension onto SPCE followed by immersing the graphite/SPCE into the electrochemical cell containing 0.1 M PBS and KCl solution and applying a constant potential of 2.0 V for 300 s. then AuNPs were electrodeposited on the surface of graphite/SPCE. This sensor provided an excellent selectivity even in 1000 fold concentrations of common interfering metal ions such as Cl^-, Br^-, SO_4^{2-}, Na^+, K^+, Ca^{2+}, Ni^{2+} and Zn^{2+}.

On the other hand, AuNPs modified CPE were used for the determination of paracetamol, neurotransmitters, and morphine by Atta et al. [24, 25]. Further modification of AuNPs/CPE with nafion was achieved also by the same group [26, 27]. The presence of nafion enhances the sensitivity of AuNPs/CPE toward the studied molecules.

Yadav et al. [28] used the combination of AuNPs and graphite in sensing application but this time in the presence of poly-1,5-diaminonaphthalene(pDAN) for the determination of Cefpodoxime Proxetil in the biological fluid. The calculated detection limit of Cefpodoxime Proxetil over AuNPs/pDAN/graphite is 39 nM in a linear range of 0.1-12 µM with a sensitivity of 4.621 µA µM^{-1}. Cefpodoxime Proxetil has a broad spectrum anti- microbial activity against several microorganisms and it is used orally for the treatment of mild to moderate respiratory tract infections, gonorrhea and urinary tract infections and also in the treatment of skin infections, acute media otitis, pharyngitis and tonsillitis [28]. Gold nanoparticles modified graphite/Sr$_2$PdO$_3$ electrode was also used as a novel non-enzymatic glucose sensor by El-ads et al. [29]. The sensor exhibited a linear range from 100 µmol L^{-1} to 6 mmol L^{-1} with a detection limit of 10.1 µmol L^{-1}. This sensor showed no interference from ascorbic acid, uric acid, dopamine, paracetamol and chloride as well as it showed a very good long-term stability. Different nanoparticles modified graphite/Sr$_2$PdO$_3$ electrodes were used in their study for glucose determination while AuNPs showed the highest performance as shown in Table 5.1 [29].

Table 5.1. A comparison between different electrodes used for the nonenzymatic glucose determination reproduced from [29].

Electrode	E_{pa} (V)	I_{pa} (mA)	Current increase (normalized to bare graphite)	Real surface area (from redox surface cm^2)	D_{app} (Cm^2 s^{-1})
Bare graphite	-0.135	0.053	----	0.110	1.08×10^{-6}
Graphite/ Sr$_2$PdO$_3$	-0.050	0.140	2.64	0.153	7.68×10^{-6}
Graphite/Au$_{nano}$	+0.072	0.200	3.81	0.297	1.57×10^{-5}
Graphite/ Sr$_2$PdO$_3$/Au$_{nano}$	+0.056	2.080	39.6	0.824	1.69×10^{-3}
Graphite/ Sr$_2$PdO$_3$/Pt$_{nano}$	+0.003	1.270	24.2	0.409	6.32×10^{-4}
Graphite/ Sr$_2$PdO$_3$/Pd$_{nano}$	-0.052	0.458	8.72	0.3512	8.22×10^{-5}

Furthermore, the determination of diethylstilbestrol and bisphenol A was enhanced upon modification of CPE with Bi_2WO_6 nanoplates [30]. The electrochemical impedance spectroscopy showed a lower electron transfer resistance for the Bi_2WO_6-CPE compared with CPE. Additionally, Bi_2WO_6 nanoplates could accelerate the rate of electron transfer for the electrochemical reaction of diethylstilbestrol (DES) and bisphenol A (BPA) on the electrode. This sensor showed low detection limits of 15 nM (S/N=3) and 20 nM (S/N=3) for DES and BPA in the linear range of 50-2100 nM and 70-1300 nM, respectively.

5.2.3. Mediator and Organic Modifier

Transition metal complexes, as electron mediators were used for the improvement of the sensitivity and lowering the overpotential of the electrode processes [31]. Metal organic frameworks (MOFs) contain inorganic and organic networks that can be easily functionalized. MOFs have received a considerable attention due to their high specific surface areas and specific micropore volumes [32, 33]. A $Zn_4O(BDC)_3$ (MOF-5; BDC = 1,4-benzenedicarboxylate) modified carbon paste electrode was used for lead determination by Wang et al. [34]. MOF-5 has the capability to adsorb lead ions that leads to an enhancement in the electrochemical performance toward lead determination. This sensor provided a limit of detection 4.9×10^{-9} in the linear range from 1.0×10^{-8} to 1.0×10^{-6} mol L^{-1} ($R^2 = 0.9981$). The modified electrode was prepared by mixing the MOF-5 with graphite powder and mineral oil to form MOF-5 modified CPE.

Ferrocene functionalized calix [4] pyrrole was reported as a host molecule for anion recognition in organic solvents. The presence of anions induced the cathodic shift in the ferrocene/ferrocenium couple [35]. Ferrocene functionalized calix [4] pyrrole was used as an active component in carbon paste electrodes for the determination of anions in water as ion selective electrode [36].The sensor displayed a sensitivity sequence: $Br^- \gg Cl^- > F^- > H_2PO^-$. The limit of detection for F^- and H_2PO^- was found to be $< 1.0 \times 10^{-7}$ M.

Phthalocyanine derivatives have a similar structure to porphyrin, the functions of metal Phthalocyanines (MPCs) are based on the enhancement of the electron transfer reactions due to their 18π-electron conjugated ring system found in their molecular structures [37]. Due to their versatility, high catalytic activity, and low cost of raw materials, MPCs attract a great of attention in wide range of applications [38]. Iron (II) phthalocyanine and its complexes are of special interest because of their high catalytic activities in various chemical and electrochemical reactions. Iron(II) phthalocyanine was used to modify CPE and used as an electrochemical sensor for the determination of epinephrine (EN), ascorbic acid (AA) and uric acid (UA) [31]. The presence of iron (II) phthalocyanine leads to an improvement of the reversibility of the electrode response and lowering its overpotential by more than 100 mV. A linear range of 1–300 μM, with a detection limit of 0.5 μM was obtained for EP determination. Another phthalocyanine sensor based on gold nanoparticles/cobaltphthalocyanine modified carbon paste electrode (AuCoPcCPE) was constructed for the determination of Dubutamine by Atta et al. [37]. The proposed sensor exhibited a low detection limit of 0.84×10^{-7} mol L^{-1} in the linear dynamic range of 6.0×10^{-6} mol L^{-1} to 2.0×10^{-4} mol L^{-1} measured in human urine. AuNPs deposited several metal phthalocyanines (Co, Ni, Cu, and Fe) modified CPE were also studied by the same group for morphine determination in the presence of ascorbic acid and uric acid. [39] Au-CoPcM-CPE showed the highest performance and the highest charge transfer rate. A lower limit of detection of 5.48×10^{-9} mol L^{-1} in the range of 4.0×10^{-7} to 9.0×10^{-4} mol L^{-1} was obtained for Au-CoPcM-CPE. Fig. 5.2 showed a comparison between metal phthalocyanines used for the electrochemical determination of morphine. The authors claimed that "the enhancement of the charge transfer rate at the electrode/electrolyte interface is due to the good synergism of metal phthalocyanine mediator which increases the electronic conduction of the paste and the gold nanoparticles which act as tiny conduction centers at the electrode surface" [39].

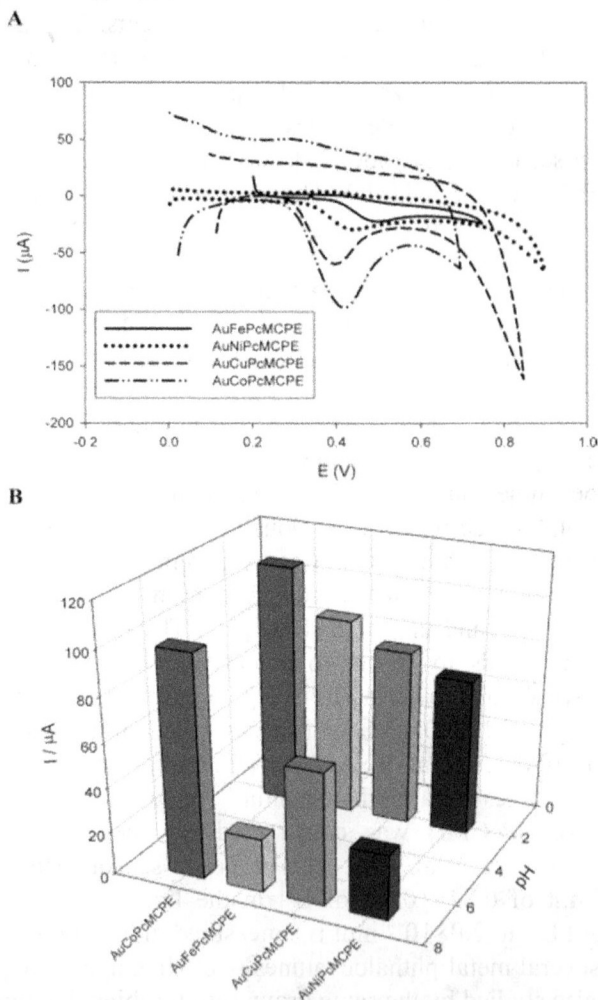

Fig. 5.2. A) Cyclic voltammograms of 1.0×10^3 molL^{-1} MO in BR buffer pH 7.4 at scan rate 100 mV s^{-1} recorded at four different working electrodes: 1) AuFePcMCPE (solid line); 2) AuNiPcMCPE (dotted line); 3) AuCuPcMCPE (dashed line), and 4) AuCoPcMCPE (dashed dotted line). B) Histogram of 1 mM MO at four different electrodes (AuCoPcMCPE, AuFePcMCPE, AuCuPcMCPE and AuNiPcMCPE) in BR buffer of pH 2 and pH 7 (reused with permission of [39]).

N,N-dimethylcyanodiaza-18-crown-6 (DMCDA18C6) is an excellent
ionophore thus, it can be used for the determination of heavy metal ions
[40]. An ultra-trace amount of lead was determined by Ganjali using a
poly vinyl chloride (PVC) membrane containing DMCDA18C6 and
coated on graphite electrode [40]. The sensor exhibits a very wide
concentration range for Pb^{2+} ions determination (from 1.0×10^{-2} to
1.0×10^{-7} M) with a limit of detection 7.0×10^{-8} M (~ 14.5 ppb). It is also
showed a fast response time of ~ 10 s and long-term stability. The sensor
also can be applied for the direct measurement of Pb^{2+} in edible oil,
human hair and water samples.

Additionally, dicyclohexyl-18-crown-6 modified carbon paste electrode
(DCH-MCPE) was used for the selective determination of Tl(I) by
Cheraghi [41]. The sensor exhibited a good reproducibility and a linear
range from 3.0 to 250 ng mL^{-1} with a limit of detection of 0.86 ng mL^{-1}.

5.3. Graphene

Graphene is a member of 2-dimentional materials discovered by Andre
Geim's research group at the University of Manchester in 2004 and
prepared by the so-called "scotch-tape" technique [42]. It composed of
hexagonal sp^2 carbon atoms aligned in a honey comb structure [43-45].
It can be considered as the mother of all carbon materials. Graphite is
multilayer graphene sheets stacked together. Single-walled carbon
nanotubes (SWCNTs) are considered as rolling up graphene sheet
[46, 47]. In addition, graphene can be coalesced to form fullerene [48].
Graphene has some unique properties that make it one of the most
interesting materials. The optical and electrical properties of reduced
graphene oxide are greatly affected by the spatial distribution of the
functional groups and structural defects [49]. It has a large theoretical
surface area about 2620 m^2 g^{-1} [47, 50, 51], has high chemical stability
and almost impermeable to gasses, withstand large current densities, has
high thermal and chemical conductivity [52, 53] in addition to its cheap
production cost compared to CNTs [53]. It also has high charge carriers,
which behave as massless relativistic particles (Dirac fermions) [54, 55].
At these Dirac points, the valence and conduction bands are degenerated
making graphene a zero band gap semiconductor. These unique
properties of graphene make it promising for wide range of applications
in the electrochemical field [52] such as Electrocatalysis [47, 51, 62],
electronics [63], organic light-emitting diodes (OLEDs), energy storage

and conversion (supercapacitors [64], batteries [65], fuel cells [66-71] and solar cells [63, 72]), sensors and biosensors [73-75]. Moreover, it represents a promising catalyst carrier in the next generation of carbon-based supports [53]. Several typical methods have been developed for graphene preparation such as "peeling-off" highly oriented pyrolytic graphite (HOPG) [56], chemical vapor deposition (CVD) [57], epitaxial growth on an electrically insulating surface [58], unzipping of CNTs [59,60], arc-discharge technique [60, 61] micromechanical exfoliation of graphite [42, 46], solvothermal synthesis [58], and thermal, chemical and electrochemical of graphene oxide (GO).

5.3.1. Graphene Functionalization

Due to the inertness as well as zero band gap structure of graphene, its functionalization attracts a significant interest. The electronic, electrical as well as the chemical properties of graphene can be tuned by either covalent or non-covalent functionalization. Non-covalent functionalization of graphene is widely discussed and developed during the last years. It includes non-covalent attachment of metal or metal oxides nanoparticles [76-78], polymers [79-82], ionic liquids [82, 83], etc. While little work is focused on the covalent functionalization of graphene, the scope of this part of the chapter is highlighting the covalent functionalization of graphene such as hydroxylation, hydrogenation, halogenation, substitutional doping and the covalent addition of organic groups and polymers. The applications of the covalently functionalized graphene in sensing and biosensing area will be covered in the next paragraphs.

The covalent functionalization of graphene sheets with organic species attracts a great of interest and has been developed for several purposes. Enhancing the solubility is one of the purposes of the covalent functionalization of graphene as well as it offers new properties differ from the properties of both pristine graphene and the organic compound. When the organic molecules are attached to graphene surface, the aromatic characters are perturbed enabling a control of the electronic properties of graphene. It is worth to mention that, the dispersion of graphene sheets in a suitable organic solvent is a crucial step in the functionalization process [84].

5.3.2. Hydrogenation and Halogenation of Graphene

Exposure of graphene to cold hydrogen plasma results in graphene hydrogenation forming what is known by graphane. Sofo et al [85] theoretically predicted the fully hydrogenated graphene (graphane) for the first time. As a result of graphene hydrogenation, the sp^2 hybridized carbon atom changed to sp^3 that leads to a significant change in its electronic properties. The charge carrier mobility of graphane is 3 orders of magnitude smaller than that of graphene so, graphane behaves as an insulator [86, 87]. As well as, C−C−C angle in graphane decreases from 120° (in graphene) to 109.5° and the C−C bond length increases from 1.42 Å (in graphene) to 1.52−1.56 Å in graphane [88, 89]. Graphane shows two additional Raman peaks at 1620 cm^{-1} (D') and 2D (2680 cm^{-1}) besides the typical G (1580 cm^{-1}) and 1350 cm^{-1} (D, the defect peak) Raman peaks of graphene [86]. It is worth also to mention that the hydrogenation process of graphene is reversible and graphane can be returned back to graphene by annealing at 450 °C in an Ar atmosphere [86]. So, we can conclude that the properties of graphene especially the electronic and electrical properties can be tuned by the partial hydrogenation of graphene. Furthermore, graphane has a ferromagnetic property and tunable band gaps depending on the hydrogenation extent [90]. Graphane can undergo a reversible hydrogenation so it is a good candidate for hydrogen storage applications [91].

The reaction of graphene with XeF_2 results in fluorination of graphene to form fluorographene or graphene fluoride [92]. This reaction is usually performed for CVD graphene. The structure of the resulted material is based on the substrate on it graphene is grown. If Cu substrate is used, the fluorination will take place on only one side and leads to C_4F stoichiometry. While C_1F_1 stoichiometry was obtained if graphene was grown on Si substrate where the fluorination takes place on both sides. Raman spectra of the fluorinated graphene from one side showed a D peak at 1350 cm^{-1} and broadening of the G (1580 cm^{-1}) and D` (1620 cm^{-1}) peaks. While fluorination of graphene on both sides showed no Raman peaks [92]. Another method to prepare fluorographene is by the chemical exfoliation of graphite fluoride [93]. This can be performed by fluorination of high-grade graphite at 600 °C in the presence of 1 atm of Fluorine gas. This process leads to the formation of $C_{0.7}F_1$ stoichiometry [93]. It is worth to mention that fluorographene is more stable than graphane [89]. The other halogens can react with graphene to form the corresponding graphene halides while their stability decreased from F to I as F>CL>Br>I [89].

The applications of hydrogenated graphene in sensors are slightly rare but Jankovsky et al. [94] prepared partially hydrogenated graphene and tested it for oxygen reduction reaction and Picric acid determination. They initially prepared GO by Hummer method (used potassium permanganate) and Hoffman method (used potassium chlorate). The reduction step was performed in the presence of two different catalysts; Zn in acid medium and Al in basic medium. To prepare the partially hydrogenated graphene using Al as a catalyst, they ultrasonicated 100 mg of GO with 100 ml of 1.0 M NaOH for 30 min then 1.5 g Al powder was added to the suspension and stirred for 24 h. the same procedure was applied in the case of using Zn catalyst except that GO was ultrasonicated first with 100 ml water then 10 ml of concentrated HCl and 2.5 g Zn powder were added before the stirring process. They tested all samples toward electrochemical sensing of picric acid (organic explosive) and they found that the partially reduced graphene prepared by any of the mentioned methods showed a higher performance compared to GCE as illustrated in Fig. 5.3. Picric acid is water soluble and contains several electrochemically active nitro groups so it can serve as a proxy compound for sensing testing applications. The CVs showed four peaks; the first reduction peak is a four-electron process to form a hydroxylamine group and it is reversible. The other peaks are irreversible. Peak II is associated with the reduction of the para-nitro group of picric acid. Peaks III and IV are attributed to the reduction of the remaining ortho-nitro group, as well as to the reduction into an amino group [94].

In another work, Islam et al. [95] have studied theoretically the interaction of hydrogenated graphene with CH_4 vapor using first principal calculations. They studied the structural, electronic and gas sensing properties of pure, defected and light metal-doped graphane monolayer. The metal used for doping are Li, Na, and Al that have a strong binding with graphane sheet at small doping concentration.

The energetic analysis revealed the weak binding of CH_4 to the pristine graphane sheet restricting its sensing application in ambient conditions. While upon metal doping, the adsorption energies significantly increase. Regarding Al doping, Al adatom donating a significant portion of its electronic charge to the graphane sheet and attains a partial positive charge. This makes the adsorption of the CH_4 molecule stronger on Al-doped graphane compared to the others.

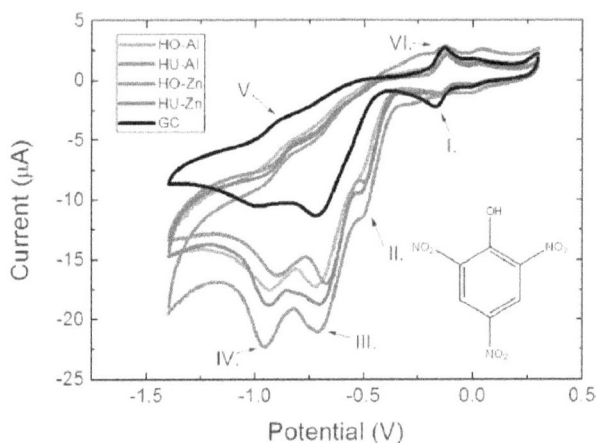

Fig. 5.3. The cyclic voltammogram of 10 mm solution of picric acid
on the reduced/hydrogenated graphene in PBS supporting electrolyte (pH 7.2).
Scan rate=100 mVs^{-1} (reused with permission of [94]).

Very recently, Zhang et al. [96] studied the plasma-fluorinated graphene
and its application as a gas sensor. In this work, fluorinated graphene
(FG) is synthesized by a controllable SF$_6$ plasma treatment and its
structure was studied via various spectroscopies, including Raman,
X-ray photoemission spectroscopy (XPS) and near-edge X-ray
absorption fine structure (NEXAFS). The angle dependent NEXAFS
reveals that the fluorine atoms interact with the graphene sheet to form
the covalent C-F bonds, which are perpendicular to the basal plane of
FG. Based on their study, FG shows much better performance for
ammonia detection compared to the pristine graphene. They attributed
the higher performance of FG to the enhanced physical absorption due
to the presence of C-F covalent bonds on the surface of FG. The
FG-based gas sensor exhibits fast response/recovery behavior and high
sensitivity of detecting ammonia from 2 to 100 ppm in 30 s at room
temperature.

5.3.3. Hydroxylation of Graphene

Recently, hydroxylation of graphene attracts a great of attention in order
to obtain a hydrophilic and biocompatible graphene [97]. One method to
obtain OH-functionalized graphene (graphene-OH) is that performed by
Yan et al. [97] using ball-milling in the solid state, they could obtain

221

graphene-OH from solid KOH and graphite powder in one step. The resulted graphene-OH showed biocompatibility to human retinal pigment epithelium cells [97]. Another method to prepare large-scale graphene-OH (kilogram scale) is that reported by Sun et al [98]. Their method is based on the initial preparation of GO followed by its dispersion in 2.0 M NaOH to form 2.0 mg mL^{-1} solution. The dispersion liquid then was sheared and emulsified by high-pressure disrupter at 207 MPa for 2 h. Graphene-OH was obtained by filtration and washing and the yield of graphene-OH is 81 % of the initial graphite used [97]. Starting by 1.0 kg graphite, 0.79 kg graphene-OH can be obtained by this method. A Digital photograph of the preparation process of G-OH paper by routine filtration technique, photographs of the flexible G-OH paper with metallic luster, FESEM image of the surface of G-OH paper and Cross-section SEM image of the G-OH paper are shown in Fig. 5.4.

In the same work, they reported that graphene-OH can be reversibly converted into Graphene-I by the direct halogenation as shown in Fig. 5.5. The resulted graphene-OH papers were directly used as effective biologically active substrates with a good mechanical property. The authors claimed that this strategy to obtain graphene-OH can be used in bone, vessel and skin regeneration as a new biomaterial [98].

NO_2 and CO_2 gas sensor was constructed by Wu et al. [99] based on 3D-chemically functionalized reduced graphene oxide hydrogel (FRGOH). The addition of hydroxyl group to RGO was achieved using hydroquinone molecules. FRGOH gas sensor displays faster recovery, lower detection limit (200 ppb for NO_2) and two-fold higher sensitivity toward both NO_2 and CO_2 compared to reduced graphene oxide hydrogel (RGOH).

5.3.4. Carboxylation and Addition of Organic Groups

The addition of organic groups to graphene sheets can be carried out mainly via two routes: i) formation of a covalent bond with sp^2 carbon atoms of graphene through an addition of dienophiles or free radicals, ii) formation of a covalent bond with oxygen groups of GO. The reaction of free radicals with graphene is usually performed through heating of diazonium salt. Tour and coworkers [84, 100] decorated graphene with nitrophenyls through free radical interaction with graphene. Once diazonium salt is heated, highly reactive free radicals are produced and attack sp^2 carbon atoms of graphene forming a covalent bond with them.

The conductivity of graphene sheets decreases as a result of the transformation from sp^2 to sp^3 hybridization. Band gap of graphene is also affected by nitrophenyls functionalization of graphene sheets as mentioned elsewhere [101]. This finding is useful in using the functionalized graphene in semiconducting applications.

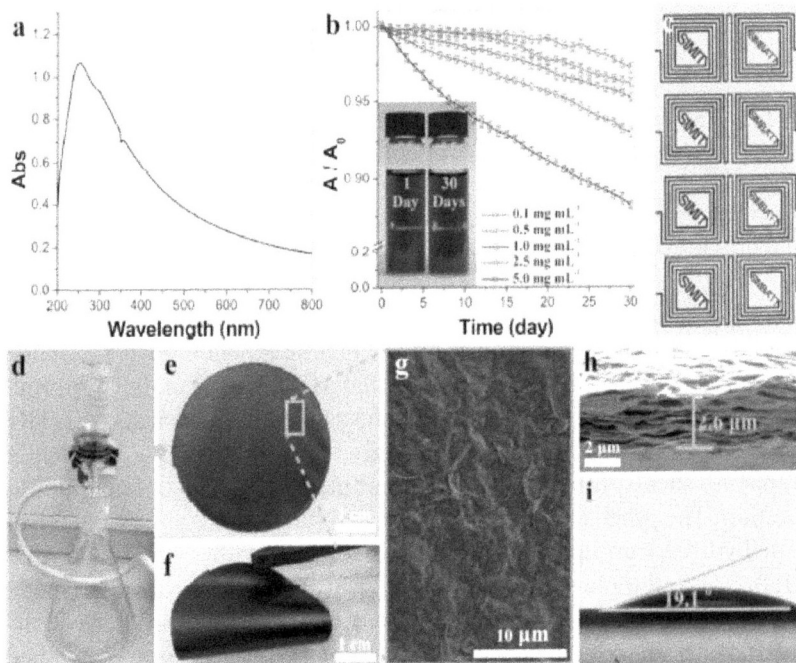

Fig. 5.4. Stable G-OH aqueous dispersion. (a) UV-vis spectrum of G-OH aqueous dispersion. (b) Sedimentation behavior of G-OH aqueous dispersion with different concentrations (0.1, 0.5, 1.0, 2.5 and 5.0 mg mL-1). Ao and A is the absorbance (at 660 nm) of G-OH aqueous dispersion stockpiled for 0 to different days. Inset is the digital photographs of 5.0 mL homogeneous dispersions (0.5 mg mL-1) stockpiled for 1 and 30 days. (c) Digital photograph of G-OH plated circuit on paper. (d) Digital photograph of the preparation process of G-OH paper by routine filtration technique. (e, f) Photographs of the flexible G-OH paper with metallic cluster. (g) FESEM image of the surface of G-OH paper. (h) Cross section SEM image of the G-OH paper showing the uniform thickness (2.6 μm) and layered structure. (I Contact angle of G-OH paper for water (reused with permission of 98).

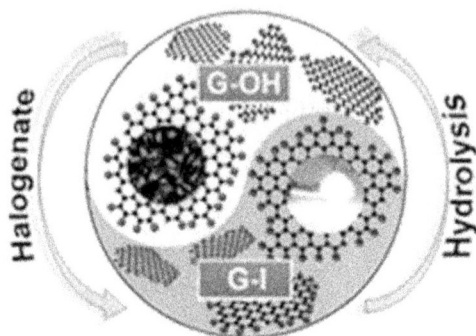

Fig. 5.5. The reversible transformation of graphene-OH into graphene-I
(reused with permission of 98).

Diazonium salts have been used to functionalize the chemically or thermally converted graphenes, single graphene sheets obtained by micromechanical cleavage of graphite, and epitaxial graphenes. [84, 102-105]. The free radical attack is another method for graphene functionalization with a polymeric material. This can be performed through the combination of atom transfer radical polymerization (ATRP) with diazonium addition reaction. Fang et al [106] added polystyrene to graphene sheets through ATRP combined with diazonium addition reaction. The process is based on the initial functionalization of graphene with hydroxyl groups through heating of GO with hydrazine hydrate (HH), 2(4-aminophnyl) ethanol and isoamyl nitrate. The second step is adding the initiator to graphene-OH forming graphene-initiator which is the starting material of polymerization of styrene on graphene sheet. Fig. 5.6 shows the preparation procedure of graphene-polystyrene. Additionally, benzoyl peroxide reacts also with graphene sheets by free radical addition method. Graphene was deposited on a silicon substrate and immersed in benzoyl peroxide dissolved in toluene. An Ar-ion laser beam was focused onto the graphene sheets in the solution to initiate the reaction. This reaction leads to adding a phenyl group to graphene sheet that results in the appearance of strong D band at 1343 cm^{-1} due to the formation of sp^3 carbon atoms [107].

Liu et al [107] explained the mechanism of this reaction and they mentioned that the hot electron initiates the photoexcited graphene and results in an electron transfer from graphene to the physisorbed benzoyl peroxide. Benzoyl peroxide radicals have a very short lifetime and rapidly decompose to phenyl radicals, which in turn react with graphene sheets forming graphene functionalized with phenyl group [107].

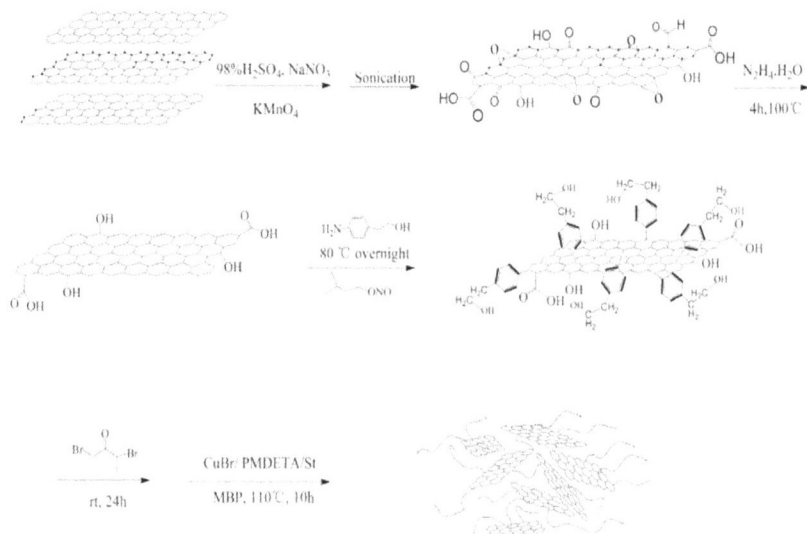

Fig. 5.6. Synthesis of polystyrene-functionalized graphene nanosheet (reused with permission of [106]).

The addition of dienophiles to graphene is carried out through 1,3 dipolar cycloaddition [108]. The most common dienophiles used to functionalize carbon materials is Azomethine ylide which provides a variety of organic derivatives functionalization and consequently a wide range of applications [109, 110]. Meanwhile, Georgakilas et al [111] used 1,3 cycloaddition of Azomethine addition to functionalize graphene sheets by pyrrolidine ring. The azomethine ylide was prepared by condensation of 3,4-dihydroxybenzaldehyde and sarcosine. Azomethine ylide addition to graphene sheets results in increasing the dispersibility of graphene in polar solvents [111]. 1,3 cycloaddition to graphene was also used to add tetraphenylporphyrin and its Pd analog to graphene sheets by Zhang and coworkers [112]. On the other hand, the reactions of graphene with organic azides such as alkylazides and azidophenylalanine were successfully carried out by several research groups [113, 114]. The resulted composites showed an enhanced dispersibility in many common organic solvents such as acetone and toluene [114]. He and Gao [115] developed a new and facile method to functionalize graphene sheets with various functional groups and polymeric chains using nitrene cycloaddition. The resulting functionalized graphene can be easily modified through different chemical reactions such as amidation, surface-initiated polymerization, and reduction of metal ions.

The other rout to covalently functionalize graphene sheets with organic group is through starting with GO and forming a covalent bond with its oxygenated groups. GO has hydroxyl, carboxylic and epoxy groups that help in the covalent attachment with various organic compounds. For example, amine-terminated oligothiophene makes amide bond with GO that is confirmed by FT-IR (Fourier transform infrared) and the resulted composite showed a superior optical limiting effect [116]. GO can be functionalized by CH_2OH terminated poly(3- hexylthiophene) (P3HT) through the formation of ester bonds as reported by Yu et al [117]. Porphyrins, phthalocyanines, and azobenzene can be also covalently attached to graphene starting with GO [118, 119]. Porphyrins attachment is based on the formation of an amide bond between carboxylic groups of GO and amine group of amine-functionalized porphyrins [120]. Covalently attached poly methylmethacrylate to GO was performed by Pramoda [121]. The procedure involved the initial functionalization of GO with Octadecylamine (ODA) that reacts with methacryloyl chloride to incorporate polymerizable –C=C–functionalities to graphene surfaces. Imidazolium derivatives can be also attached to GO through its epoxy group as presented by Yang et al [122]. The imidazolium-modified GO was easily dispersed in water, N,N-dimethylformamide, and dimethyl sulfoxide. Karousis et al. [123] reported that 1-(3-aminopropyl)-imidazole can attach to GO through its carboxylic group.

Graphene oxide (AGO) functionalized with amid group was used as a humidity sensor as in [124]. The amidation of GO was obtained by condensation of amine group of heteroaryl/phenylamine with lactone group of GO. The procedure involves ultrasonication of 0.3 g GO in 30 ml DMF for an hour before adding 0.3 g NaOH then the mixture was stirred for an hour. 2-aminothiazole (3.1 mmol), hydroxybenzotriazole (HOBt) (3.1 mmol) were added followed by 2-aminothiazole, N,N'-dicyclohexylcarbodiimide (DCC) (3.1 mmol) and stirred for 24 hours at room temperature. The resulted AGO showed much better response at the lower and the higher relative humidity (10 % to 90 %) indicates its effectiveness compared to GO.

The direct electrochemistry and electrocatalysis of catalase (Cat) were studied by Huang et al. [125] using AuNPs/ graphene-NH$_2$. Cat/AuNP/ graphene-NH$_2$/GCE modified electrode was used as a biosensor for H$_2$O$_2$ determination. Cat/AuNP/ graphene-NH$_2$/GCE biosensor showed a fast response time of (2 s) and a linear range from 0.3 to 600 μM with a detection limit of 50 nM at S/N = 3 by amperometric technique. The preparation of amine functionalized graphene involves two steps. Initially carboxylic acid functionalized graphene was prepared by placing GO that washed several times with HCl and H$_2$O into a quartz tube that was sealed at one end and the other end was closed using a rubber stopper. The sample was flushed with argon for 10 min through the rubber stopper, and the quartz tube was inserted into a Lindberg tube furnace preheated to 1050 °C and held in the furnace for 30 s. the final step is the amination of graphene by chlorination of graphene-COOH through refluxing for 12 h with SOCl$_2$ at 70 °C. and then reaction with NH$_2$(CH$_2$)$_2$NH$_2$ in dehydrated toluene for 24 h at 70 °C. the electrode surface was prepared as illustrated in Fig. 5.7 and used directly for H$_2$O$_2$ determination. The biosensor showed also very good cyclic and long-term stability where the biosensor showed no decrease in the current after 100 cycles and only 4.2 % decrease after 15 days stored in the refrigerator [125].

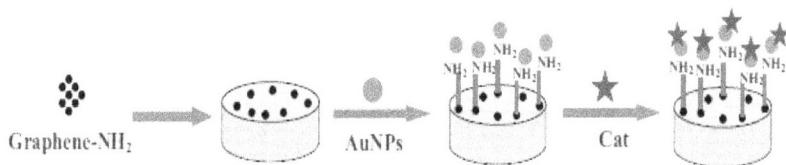

Fig. 5.7. Schematic diagram showing the fabrication of Cat/AuNPs/graphene-NH$_2$/GCE (© 2012 ul Hasan K, et al. [125]).

Hasan et al. [126] developed another glucose sensor based on peptide bonds between the amine groups of GOD and the residual carboxylic functionalities of GO. GO was functionalized initially by polyquaterniums, a kind of polymeric ionic liquid (PIL). Polyquaterniums have a positive charge and are widely used in hair shampoos and conditioners to neutralize their negative charge. The same idea was used in their paper; polyquaterniums are used to neutralize the negative charge of GOD hence, results in an efficient glucose sensing device. A schematic of the GOD immobilization on graphene showing the peptide bond formation is depicted in Fig. 5.8. To construct the biosensor, GO was mixed with PIL and reduced by hydrazine hydrate then graphene was coated on a hot Pt wire by the dip coating. GOD enzyme was electrostatically immobilized by dipping the graphene-coated Pt wire into the enzyme solution for 15 minutes. The resulting biosensor showed a broad linear range up to 100 mM glucose concentration and with a sensitivity of 5.59 µA/ decade [126].

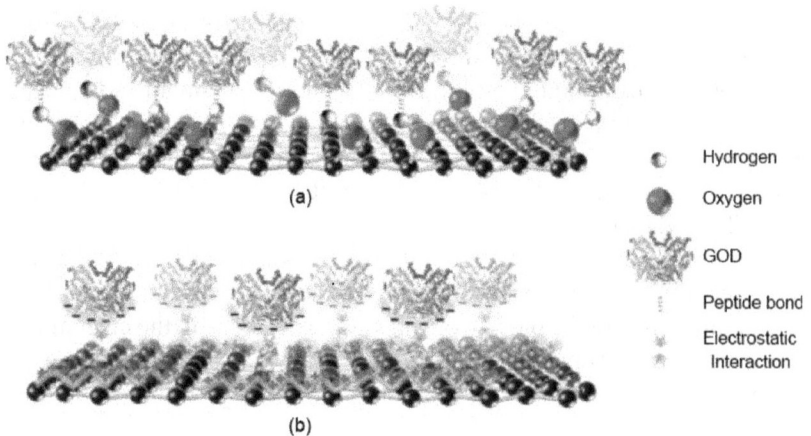

Fig. 5.8. Schematic of the GOD immobilization on graphene: (a) Peptide bond formation between remaining carboxylic groups of RGO and GOD; (b) Electrostatic interaction between PIL functionalized graphene surface and GOD (reused with permission of [126]).

A free-standing sulfonic acid functionalized graphene oxide (fSGO)-based electrolyte film was prepared and used as alcohol fuel cell sensor by Jiang et al. [127]. FSGO showed a high sensitivity toward ethanol vapor with recorded detection limit of 25 ppm. For FSGO preparation 40 mg of freeze-dried GO was added to 40 ml toluene and sonicated for an hour. 80 mg of MPTMS were then added dropwise as a source of sulfonic acid function group. Then the solution left for 48 h in order to graph mercapto groups to GO. The oxidation reaction was carried out by adding 40 ml H_2O_2 (30 %) and leaving the reaction to complete for 24 h.

The effect of the combination of the chemical functionalization and the mechanical strain of graphene on the gas sensing performance was

studied by Bissett et al. [128]. The aryl diazonium molecules used to alter
the electronic structure of graphene [129, 130.] while the role of
mechanical strain is to introduce a significant band gap into the graphene
[131]. To prepare the covalently functionalized graphene, the graphene
resulted from CVD (on Cu foil substrate) was submerged into a solution
of 2.0 mM 4-nitrobenzenediazonium tetrafluoroborate. In order to
introduce a mechanical strain, the functionalized CVD-grown graphene
was spin-coated with polydimethylsiloxane (PDMS) then Cu foil was
etched using aqueous 1.0 M $FeCl_3$ leaving graphene/PDMS substrate that
was then strained using a custom uniaxial strain regime. They concluded
from their study that the use of 4-NBD was shown to produce p-type
doped graphene [131]. This finding indicates that the resulted chemically
functionalized graphene with a mechanical strain is a powerful candidate
to be used as a gas sensor.

Huang et al. [132] constructed a biosensor for the determination of
adenine and guanine using graphene-COOH modified GCE. The
biosensor provided good peak separation between guanine and adenine
of 334 mV as well as a limit of detection of 2.5×10^{-8} and 5.0×10^{-8} M for
the individual determination of adenine and guanine, respectively. The
determination method is based on the enhancement of the sensitivity
through electrostatic interaction between the negatively charged
graphene-COOH and the positively charged adenine and guanine. In
their work, graphene-NH_2 was also prepared for comparison through the
chlorination of graphene-COOH by $SOCl_2$ followed by amidation
through reaction with $NH_2(CH_2)_2NH_2$. Graphene-COOH showed the
highest response compared to both graphene and graphene-NH_2. But
graphene-NH_2 showed the lowest performance due to the electrostatic
repulsion.

Another way for graphene carboxylation is that performed by Hu et al.
[133] through the reaction with 3,4,9, 10-perylene tetracarboxylic acid
(PTCA). The PTCA molecule has COOH groups and a π-conjugated
structure that can interact with the basal plane of graphene sheets through
π- π interaction and hydrophobic forces [133]. So, the interaction of
PTCA with graphene sheets helps in their separation and decoration with
more –COOH groups. The resulting PTCA/graphene composites
disperse well in many solvents. This also provides active sites for the
immobilization of the 5`-NH_2 modified probe DNA sequence. This
method was used as a label-free electrochemical method for DNA
hybridization detection by EIS measurement. ssDNA lies on the

graphene basal plane via p-stacking attractions between the bases of ssDNA and the hexagonal cells of graphene. The "lying" ssDNA after hybridization with its complementary sequence became "standing" dsDNA. They monitored the Negative-charge change and conformation transition upon DNA immobilization and hybridization using EIS and adopted as the hybridization signal. So, their method is considered as a label-free strategy for the HIV-1 pol gene sequence detection [133].

5.3.5. Substitutional Doping with Foreign Atoms

In the substitutional doping, carbon atoms from the hexagonal lattice are substituted by foreign atoms that are usually nitrogen or boron atoms. This process leads to the formation of n- or p- type semiconductor behavior of graphene depending on the substituted atom. Substitution with nitrogen leads to an increase in the electron density and hence, graphene with n-type characteristics is formed. On the other hand, boron substitution leads to the formation of p-type graphene due to holes formation upon boron substitution [134]. There are several methods to incorporate foreign atoms to graphene skeleton for example Wang et al. [135] prepared N-doped graphene using a high-power electrical heating in an ammonia gas. The same group of research prepared N-doped graphene by the simultaneous doping and reduction of GO into RGO by annealing GO in the presence of NH_3 gas [136]. The maximum N-substitution was 5 % and achieved at 500 °C. They also observed that NH_3 is more effective to reduce GO than H_2 gas under their conditions of preparation. Hydrothermal treatment of GO was also used to prepare N-doped graphene [137] Wei et al showed that a high percent of N-substitution, 8.9 %, was achieved using CVD technique in the presence of NH_3 gas [138]. The arc-discharge method was also used to prepare B-doped graphene by using a mixture of hydrogen and diborane vapor or by using boron-stuffed graphite [139]. The solvothermal method is another method to prepare N-doped graphene. This method provides a high content N-doping reaches 16.4 % and leads to formation of N-doped graphene in the gram scale [140]. Sheng et al. [141] used N-doped graphene (NG) for the simultaneous electrochemical determination of dopamine (DA), AA and UA. NG showed a detection limit of 2.2×10^{-6} M, 2.5×10^{-7} M and 4.5×10^{-8} M for AA, DA and UA in the linear range of 5.0×10^{-6} to 1.3×10^{-3} M, 5.0×10^{-7} to 1.7×10^{-4} M and 1.0×10^{-7} to 2.0×10^{-5} M, respectively. For the simultaneous determination, the electrode material showed a linear response for AA, DA and UA with a good peak separation. The corresponding linear ranges for AA, DA and UA determination are 1.0×10^{-5} to 6×10^{-4} M,

1.0×10^{-6} to 1.4×10^{-4} M and 2.0×10^{-6} to 1.6×10^{-4} M with detection limit of 3.5×10^{-6} M, 2.8×10^{-7} M and 5.7×10^{-7} M. While Feng et al. [142] could prepare a three-dimensional N-doped graphene that showed a detection limit of 1 nM toward DA and a linear range from 3.0×10^{-6} to 1.0×10^{-4} M.

Fan et al. [143] constructed a sensor for the determination of Bisphenol A (BPA) based on N-doped graphene and chitosan (CS). They stated that "N-GS has favorable electron transfer ability and electrocatalytic property, which could enhance the response signal towards BPA. CS also exhibits excellent film formation ability and improves the electrochemical behavior of N-GS modified electrode" [143]. They could achieve a detection limit of 5.0×10^{-9} M and a linear range from 1.0×10^{-8} to 1.3×10^{-6}.

N-doped graphene decorated with Ag nanoparticle was used as an electrochemical sensor for glucose determination [144]. The sensor showed fast response time (<10 s) and a detection limit of 0.02 mM.

On the other hand, doping of graphene sheets with boron leads to the formation of P-type semiconductor that is used mainly in gas sensing applications. Lv et al. [145] developed an ultrasensitive gas sensor based on B-doped graphene. The sensor used for the determination of NO_2 and NH_3. Theoretical calculation demonstrated that boron-doped graphene could break the symmetry of spin-up and spin-down transmittance channel, thus leading to a metallic-to semiconductor transition [145]. B-doped graphene was prepared by CVD technique, in which a triethylborane (TEB) was used as boron source and hexane solution as a carbon source. As shown in Fig. 5.9 BG showed enhanced performance toward NO_2 and NH_3 determination compared to the pristine graphene.

Recently, sulfur and phosphorus were used for the substitutional doping of graphene [146]. Sulfur doping is quite different from N and B doping. That is because S atom is much larger than C atom and also the difference in the electronegativity between S (2.58) and C (2.55) is very small. These make the charge transfer in C-S composites is very small [147]. The theoretical calculation showed that the structure of graphene can be distorted by S-doping and open the band gap of graphene [148]. S-doping of graphene also modifies the local chemical reactivity and showed very good performance as a gas sensor for polluting gasses like NO and NO_2 [149]. S doping of graphene can be obtained by CVD using hexane as organic source of C and sulfur powder as in [146]. The S-doped graphene preparation was confirmed by Raman analysis. S-doped graphene showed a chemical shift in both D and G band as well as in the relative intensity between them. G/D ratio is much higher in the case of S-doped graphene compared to the pristine graphene. XPS also showed a band at 163.5 which as assigned for C-S bond and did not show any peaks higher than 165 that confirms the presence of sulfur as a sulfur atom and not as oxides such as SO_3^{2-} or SO_4^{2-}. S-doped graphene was used for gas sensing applications and showed a high selectivity toward NO_2 gas compared with NH_3, CH_4, SO_2 and CO [150].

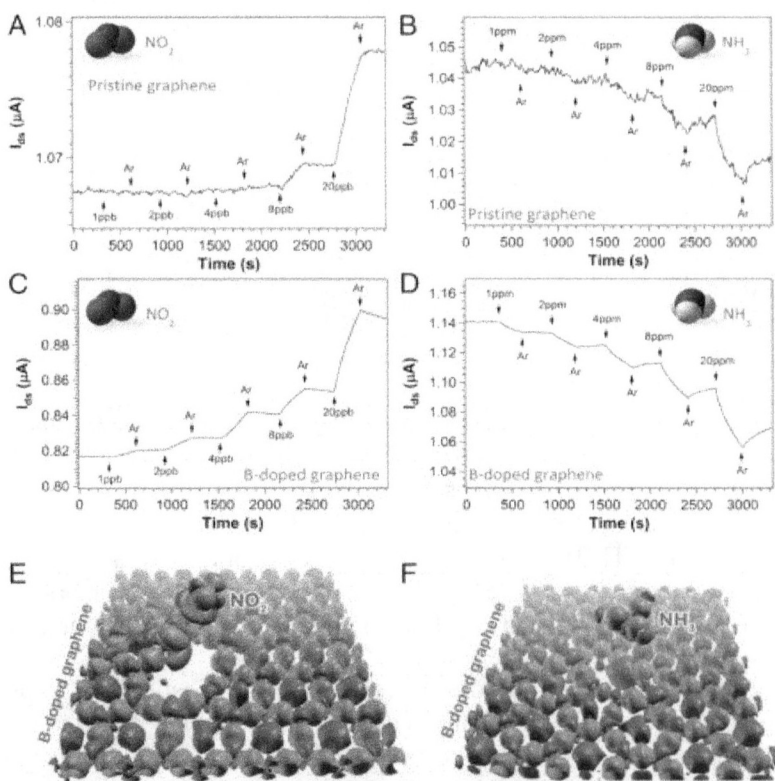

Fig. 5.9. Comparison of sensor response between PG and BG sheets. (A and B) sensor response of PG sheets versus time recorded with the sensor exposed to NO_2 (A) and NH_3 (B). (C and D) corresponding gas sensing on BG sheets when the sensor was exposed to NO_2 (C) and NH_3 (D). E and F demonstrate the difference of charge density with respect to the isolated atoms for B_3-doped graphene with NO_2 and NH_3 molecules, respectively (reused with permission of [145]).

On the other hand, Liu et al. [151] prepared S-doped graphene by microwave assisted solvothermal reaction. This method represents a fast and simple method for S-doping of graphene using benzoyl disulfide as a source of sulfur. other methods for S-doping are the thermal reaction of GO-mesoporous silica sheets in H_2S gas [152]; thermal exfoliating of GO in SO_2, H_2S, or CS_2 gas [153] and thermal shock/quench annealing of GO and phenyl disulfide [154]. S-doped graphene showed a high sensitivity and selectivity toward H_2O_2 compared to graphene. They claimed that the high sensitivity of S-doped graphene is attributed to the improved conductivity/charge transfer and creation of more active sites upon doping with S atoms. S-doped graphene showed a detection limit of 0.7 μ, a linear range from 0.1 to 18 mM and a sensitivity of 285.6 $\mu A \ mM^{-1}$.

Phosphorous atom is more electron-rich compared to a carbon atom. So, upon phosphorous incorporation to graphene matrix, its chemical and electrical properties are altered. Niu et al. [155] prepared P-doped graphene by using thermal decomposition through annealing of GO with triphenylphosphine (TPP). The preparation method involves mixing of 50.0 mg GO with 50.0 mg TPP in DMF/H_2O (1:10) solution and allow them to be annealed at high temperature in an Ar atmosphere. The as-prepared material is considered as P-type semiconductor and was used for NH_3 sensing at room temperature with enhanced activity compared to the pristine graphene. They also found that P-doped graphene that was prepared at a lower temperature (400 °C) showed the highest response compared to that prepared at a higher temperature. They attributed this finding to the removal of P atoms at a higher temperature which consequently affects the electrochemical performance toward NH_3 gas molecules.

On the other hand, P-doped graphene has been prepared by the electrochemical method by Thirumal et al. [156]. Their method involves an electrochemical exfoliation/erosion/expansion process with phosphate/oxygen group functionalization on graphene. Phosphoric acid H_3PO_4 was used for graphene exfoliation and as a source of phosphorous for P-graphene preparation. A constant 5.0 V was subjected to two graphite electrodes immersed in H_3PO_4 for 5.0 hours in order to prepare P-doped graphene. The black suspension was collected at the end of the experiment and centrifuged several times for its washing.

Dual doping has received a great of interest and showed a great enhancement in the electrochemical sensing performance of graphene [157]. The enhanced performance is attributed to the synergistic effect arising from the coupling interactions between two heteroatoms [157]. Chen et al. [158] have prepared S-N dual doped graphene (NS-G) through two-step solvothermal process and used it as a matrix for GOD immobilization for glucose determination. The prepared NS-G showed a better glucose sensing performances in terms of the detection sensitivity, limit of detection, and linear range compared to the biosensor based on single-doped N-G. NS-G was synthesized by a simple two-step solvothermal method using urea as the N source and benzyl disulfide (BD) as the S source. Firstly, N-G was prepared through ultrasonication of 200 mg of GO dispersed in 500 mL of deionized water for 1 h. Then 2 g of urea was added into the GO suspension and ultrasonicated for another 1 h. The suspension was sealed in a Teflon-lined autoclave and treated at 170 °C for 8 h. The composite was cooled down and was collected by centrifugation. 50 mg of N-G was mixed with 0.5 g of BD and the process was repeated but with using solvothermal conditions of 190 °C for 10 h. For the electrochemical determination of glucose, 5.0 mg of NS-G was added to 2 mL of chitosan solution (0.5 wt.%, 2.0 % acetic acid) and ultrasonicated for 1.0 h. Next, 5μL of the above suspension was cast onto the surface of glassy carbon electrode (GCE) and dried in air to form NS-G/GCE. Then the modified NS-G/GCE was immersed in 0.1 M PBS (pH = 7.4) containing GOD (4 mg/mL) at 4 °C for 24 h in order to immobilize GOD on the electrode surface and obtain the GOD/NS-G/GCE. The sensor showed a wide linear range of 0.01-17.2 mM with a low detection limit of 0.6 μM and a high selectivity of 145.7 $\mu A \cdot mM^{-1} \cdot cm^{-2}$. Yang et al. [159] prepared B, N dual-doped graphene modified sensor that showed higher performance than the single doped (B or N) graphene. The sensor used for the electrochemical detection of H_2O_2. Guo et al. [160] fabricated N, S dual-doped graphene modified sensor for the simultaneous determination of Pb^{2+} and Cd^{2+} ions.

On the other hand, Choudhuri et al [161] used the theoretical calculations by density functional theory (DFT) to investigate the using of B-N doped graphene for gas sensing applications. The calculations showed that the electronic properties of the B-N@graphene surfaces change significantly compared to pure, B@ and N@graphene surfaces. They also reported that B-N co-doping on graphene can be highly sensitive and selective for semiconductor-based gas sensor compared to the single doped graphene.

5.4. Combined Carbon Materials

Very recently, many researchers have considered the using of mixed
carbon materials as an electrode surface to be used in many applications.
One of the reasons for this combination is to increase the surface
roughness of the carbon material in order to get a higher surface area.

Carbon-paste electrodes (CPEs) are widely used in the electrochemical
determinations of various biological and pharmaceutical species due to
their low residual current and noise, wide anodic and cathodic potential
ranges, rapid surface renewal, ease of fabrication and low cost.
Moreover, it is easy to modify CPE by adding different substances into
it in order to increase the sensitivity, selectivity, and rapidity of the
determinations [162].

Carbon nanotubes (CNTs) have been used in carbon paste electrodes for
fabricating electrochemical sensors and biosensors instead of the casting
technique on the conventional electrode surfaces. Owing to their special
physicochemical properties, CNTs attracts a considerable attention in the
sensing applications. They have an ordered structure with a high surface-
to-volume ratio, excellent mechanical strength, ultra-light weight, high
thermal conductivity, high electrical conductivity, and chemical
performance. Guo et al [163] fabricated a chemically functionalized
multi-walled CNTs (MWCNTs) and mixed them with graphite powder
and mineral oil to form MWCNTs paste electrode. The initial step was
the functionalization of MWCNT with –COOH followed by its reaction
with 2-aminothiophenol (L) to form L-grafted MWCNTs
(L-g-MWCNTs). L-g-MWCNTs paste electrode was used as an ion
selective electrode for the electrochemical determination of Pd^{2+} ions.
The sensor exhibits a linear range from 5.9×10^{-10} to 1.0×10^{-2}M with a
detection limit of 3.2×10^{-10} and a sensitivity of 29.5 ± 0.3 mV dec^{-1}. It
could determine Pb^{2+} ions concentration in environmental samples, e.g.
soils, waste waters, lead accumulator waste and black tea. Another CNT
paste electrode is that fabricated by Liu et al. [164] based on the
formation of a poly-glutamic acid (PGA) film modified carbon paste
electrode (CPE) incorporating carbon nanotubes (CNTs) for the
determination of L-tryptophan (L-Trp). Glutamic acid was electro-
polymerized on the surface of CPE incorporating CNTs. They obtained
a linear range from 5.0×10^{-8} to 1.0×10^{-4} M with a detection limit of
1.0×10^{-8}M (S/N=3) and sensitivity of 1143.79 μA mM^{-1} cm^{-2} for L-Trp
determination. Moreover, carbon paste electrode modified with

NiO/CNTs nanocomposite and (9, 10-dihydro-9, 10-ethanoanthracene-11,12-dicarboximido)-4 ethylbenzene-1, 2-diol as a mediator was used for the simultaneous determination of cysteamine (CA), nicotinamide adenine dinucleotide (NADH) and folic acid (FA) [165]. The sensor exhibited detection limits of 0.007, 0.6, and 0.9 mmol L^{-1}, for the determination of CA, NADH and FA, respectively in biological and pharmaceutical samples. The electrode was prepared by mixing 60.0 mg of DEDE, 740 mg of graphite powder, 200 mg of NiO/CNTs with 0.4 g of paraffin in a mortar and pestle. The paste was then packed into a glass tube with an electrical contact and used directly for the measurement.

Various oxide and hydroxides were used as modifiers with CNT-graphite paste for sensing applications. Nanoparticle hybrid materials of metals, metal oxides and/or hydroxides with MWCNTs are highly promising for sensing applications. Furthermore, the uniform dispersion of nanoparticles on MWCNT surface results in an improvement of the performance of the sensing materials. For example, Fe_3O_4 nanoparticles have attracted a great of interest in sensing and biosensing applications due to their good biocompatibility, strong super paramagnetic property, low toxicity, easy preparation and high adsorption ability [166, 167]. Thus, the combination between magnetite and CNTs provides an excellent electrochemical performance due to the combination of the enlarged active surface area, a strong adsorptive capability of the nanomaterial and their specific interaction ability. This combination was used as a modifier for CPE for the electrochemical determination of haloperidol (Hp) [168]. The obtained linear ranges were from 1.2×10^{-3} to 0.52 and from 6.5×10^{-4}–0.52 μmol L^{-1} with detection limits 7.02×10^{-4} and 1.33×10^{-4} μmol L^{-1} using differential pulse and square wave techniques, respectively.

Graphene and graphene oxide were also used in a combination with graphite powder in order to obtain graphene paste electrodes. The simultaneous determination of glutathione (Glu) and penicillamine (PA) was performed using CPE modified with RGO and 10,10-dimethyl-7(3,4-dihydroxyphenyl)-10,11 dihydrochromeno[4,3-b]chromene-6,8(7H,9H)-dione (DDDC) as a mediator [169]. The sensor provided a linear response for Glu in the range of 0.08–100 μM with a detection limit (based on 3sbl/m) of 0.02 μM and it is also applied for the determination of Glu and PA in the real samples.

Graphene nanosheets and an ionic liquid (n-hexyl-3-methylimidazolium hexafluoro phosphate) were incorporated into CPE for the electrochemical determination of mangiferin [170]. Ionic liquids (ILs) have high thermal stability, no volatility, high polarity, large viscosity, high intrinsic conductivity, and wide electrochemical range. So, its combination with graphene sheets in the sensing applications leads to an enhancement in the performance of the sensor [170]. Mangiferin is used as antioxidant [171] and has an antitumor activity [172], anti-HIV activity [173], immunomodulation [174], anti-inflammation [175], antidiabetic activity [176], and hepatoprotective activity [177] so, its determination with low detection limit is so crucial. Their sensor showed an improvement in the performance of the electrode compared to bare CPE. It showed a linear range from 5.0×10^{-8} to 2.0×10^{-4} M and a detection limit of 20.0 nM for mangiferin determination.

Graphene oxide was also used as a modifier for CPE for the potentiometric determination of Cu^{2+} ions [178]. A hybrid material of 2-amino-5-mercapto-1,3,4-thiiodiazole was grafted onto nanoscale graphene oxide (AMT-g-NGO) and incorporated into CPE as a neutral carrier. The electrode exhibited a range from 1.0×10^{-7} M to 1.0×10^{-1} M with a slope of 26.2 mV dec^{-1} and detection limit (4.0×10^{-8} M). Moreover, it showed a good selectivity, a long lifetime (2 months), wide applicable pH range (3.0–7.0) and fast response time (15 s). This sensor can be used as an indicator electrode for the potentiometric titration of Cu^{2+} ions with EDTA and applicable also for the determination of Cu^{2+} ions in different real samples.

The combination between CNTs and graphene attracted a significant attention as a high-performance electrochemical sensor. Graphene-Carbon Nanotube Networks were used as wearable strain sensor by Shi et al. [179]. The stretchable films of carbon nanotubes (CNTs) have attracted a great attention especially in the field of flexible electronics. However, the lack of the structural strength in CNT networks leads to deformation and failure under a high mechanical load. The presence of graphene can enhance the strength and capabilities of CNT network, especially at the nanotube joints. Additionally, it enhances their resistance to buckling and bundling under a large cyclic strain up to 20 %. In their paper, the chemical vapor deposition (CVD) was used to prepare the sensor. Graphene was deposited on copper that is previously coated by ultrathin CNT films as templates. So, graphene homogeneously fills the nanotube voids and forms a seamless hybrid, named CNT embroidered graphene (CeG) as shown in Fig. 5.10. The presence of graphene greatly improves the strength and load transfer at the nanotube joints, with maintaining the high transparency and conductivity of the CNT networks. The sensor has been applied as strain sensors to detect human finger motions with fast response and high accuracy and the signals obtained from the bent and unbent finger are shown in Fig. 5.10.

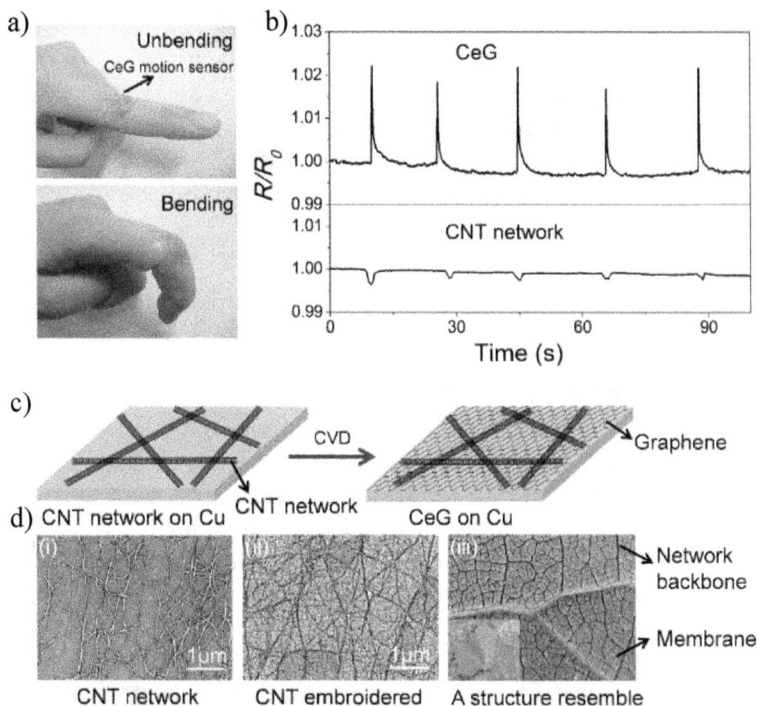

Fig. 5.10. a) The photos of the unbending/bending state of a finger, with the CeG/ultrathin PDMS adhered on the finger joint; b) The resistance recording during five bending–releasing processes using CNT network and CeG based flexible device, respectively; c) Schematic illustration of the synthesis of a CNT embroidered graphene (CeG) film by using ultrathin CNT network as template; d) SEM images of ultrathin CNT network (i) and derived CeG (ii), along with the optical images of a phoenix tree leaf (iii). Demonstration of the motion sensing application of CeG (reused with permission of [179]).

References

[1]. Brian McEnaney, structure and bonding in carbon material, in Carbon material for advanced technology, Timothy D. Burchell, *Elsevier*, 1999.

[2]. A. A. Zakhidov, W. A. de Heer, Carbon nanotube: the rout toward applications, Ray H. Baughman1, *Science,* Vol. 297, 2002, pp. 787-792.

[3]. J. B. Raoof, R. Ojani, F. Chekin, Fabrication of functionalized carbon nanotube modified glassy carbon electrode and its application for selective oxidation and voltammetric determination of cysteamine, *Journal of Electroanalytical Chemistry,* Vol. 633, 2009, pp. 187–192.

[4]. A. Gasniera, M. L. Pedanoa, F. Gutierreza, P. Labbeb, G. A. Rivasa, M. D. Rubianesa, Glassy carbon electrodes modified with a dispersion of multi-wall carbon nanotubes in dopamine-functionalized polyethylenimine: Characterization and analytical applications for nicotinamide adenine dinucleotide quantification, *Electrochimica Acta,* Vol. 71, 2012, pp. 73– 81.

[5]. H. Kuzmany, A. Kukovecz, F. Simon, M. Holzweber, Ch. Kramberger, T. Pichler, Functionalization of carbon nanotubes, *Synthetic Metals,* Vol. 141, 2004, pp. 113–122.

[6]. L. Meng, Ch. Fu, Q. Lu, Advanced technology for functionalization of carbon nanotubes, *Progress in Natural Science,* Vol. 19, 2009, pp. 801–810.

[7]. H. Xie and W. Yu, Functionalization Methods of Carbon Nanotubes and its Applications, in Carbon Nanotubes Applications on Electron Devices, Jose Mauricio Marulanda (Ed.), *InTech,* 2011.

[8]. H. O. Pierson, Handbook of carbon, graphite, diamond and fullerene: properties, processing and applications, *NOYES Publication,* New Jersey, 1993.

[9]. T., D., Louie, S. G., Mamin, H. J., Abraham, D. W., Thomson, R. E., Gam, E. and Clarke, Theory and observation of highly asymmetric atomic structure in scanning-tunneling-microscopy images of graphite, *Journal of Physical Review B,* Vol. 35, 1987, pp. 7790-7793.

[10]. R. N. Adams, Carbon Paste Electrode, *Analytical Chemistry,* Vol. 30, 1958, pp. 1576-1577.

[11]. T. Kuwana, W. H. French, Electrooxidation or Reduction of Organic Compounds into Aqueous Solutions Using Carbon Paste Electrode, *Analytical Chemistry,* Vol. 36, 1964, pp. 241-242.

[12]. E. Miland, A. J. Miranda Ordieres, P. Tunon Blanco, M. R. Symyth, C. O. Fagain, Poly (o-aminophenol) modified bienzyme carbon paste electrode for the detection of uric acid, *Talanta,* Vol. 43, 1996, pp. 785-796.

[13]. H. M. Abu-Shawish, N. A. Ghalw, F. R. Zaggout, S. M. Saadeh, A. R. Al-Dalou, A. A. Abou Assi, *Biochemical Engineering Journal,* Vol. 48, 2010, pp. 237–245.

[14]. P. Gupta, R. N. Goyal Polymelamine modified edge plane pyrolytic graphite sensor for the electrochemical assay of serotonin, *Talanta,* Vol. 120, 2014, pp. 17–22.

[15]. A. Nezhadali, L. Mehri, R. Shadmehri, Determination of benzimidazole in biological model samples using electropolymerized-molecularly imprinted

polypyrrole modified pencil graphite sensor, *Sensors and Actuators B*, Vol. 171– 172, 2012, pp. 1125– 1131.

[16]. M. B. Gholivand, M. Shamsipur, S. Dehdashtian, H. R. Rajabi, Development of a selective and sensitive voltammetric sensor for propylparaben based on a nanosized molecularly imprinted polymer–carbon paste electrode, *Materials Science and Engineering C*, Vol. 36, 2014, pp. 102–107.

[17]. S. Sadeghi, A. Motaharian, Voltammetric sensor based on carbon paste electrode modified with molecular imprinted polymer for determination of sulfadiazine in milk and human serum, *Materials Science & Engineering C*, 2013, pp. 4884-4891.

[18]. Dutduan Udomsap, Catherine Branger, Ge´rald Culioli, Pascal Dollet and Hugues Brisset, A versatile electrochemical sensing receptor based on a molecularly imprinted polymer, *Chem. Commun.*, Vol. 50, 2014, pp. 7488-7491.

[19]. N. F. Atta, A. M. Abdel-Mageed, Modern Applications of Molecularly Imprinted Materials, N. F. Atta, in Nanosensors: Materials and Technologies, *IFSA Publishing*, 2014.

[20]. L. Chen, X. Wang, W. Lu, X. Wua, J. Lia, perspectives and applications, *Chemical Society Review.*, 2016, 45, 2137-2211.

[21]. Zhou, L., Wang, Q., Sun, Q., Chen, X., Kawazoe, Y., Jena, P. Ferromagnetism in Semi hydrogenated Graphene Sheet, *Nano Lett.*, 2009, 9, pp. 3867-3870.

[22]. J. Pillay, K. I. Ozoemena, R. T. Tshikhudo, R. M. Moutloali, Monolayer-protectedclusters of gold nanoparticles: impacts of stabilizing ligands on the heteroge-neous electron transfer dynamics and voltammetric detection, *Langmuir*, Vol. 26, 2010, pp. 9061–9068.

[23]. Chelladurai Karuppiah, Selvakumar Palanisamy, Shen-Ming Chen, Sayee Kannan Ramaraj, Prakash Periakaruppan, A novel and sensitive amperometric hydrazine sensor based on goldnanoparticles decorated graphite nanosheets modified screen printedcarbon electrode, *Electrochimica Acta*, Vol. 139, 2014, pp. 157–164.

[24]. N. F. Atta, A. Galal, S. M. Azab, Electrochemical Determination of Paracetamol Using Gold Nanoparticles – Application in Tablets and Human Fluids, *International Journal Electrochemical Science*, Vol. 6, 2011, pp. 5082 – 5096.

[25]. N. F. Atta, A. Galal, S. M. Azab, Electrochemical Morphine Sensing Using Gold Nanoparticles Modified Carbon Paste Electrode, *International Journal Electrochemical Science*, Vol. 6, 2011, pp. 5066 – 5081.

[26]. N. F. Atta, A. Galal, S. M. Azab, Determination of morphine at gold nanoparticles/Nafion carbon paste modified sensor electrode, *Analyst*, Vol. 136, 2011, pp. 4682-4691.

[27]. N. F. Atta, A. Galal, F. M. Abu-Attia, S. M. Azab, Simultaneous determination of paracetamol and neurotransmitters in biological fluids using a carbon paste sensor modified with gold nanoparticles, *Journal of Material Chemistry*, Vol. 21, 2011, pp. 13015- 13024.

[28]. S. K. Yadav, B. Agrawal, R. N. Goyal, AuNPs-poly-DAN modified pyrolytic graphite sensor for the determination of Cefpodoxime Proxetil in biological fluids, *Talanta*, Vol. 108, 2013, pp. 30–37.

[29]. E. H. El-Ads, A. Galal, N. F. Atta, Electrochemistry of glucose at gold nanoparticles modified graphite/SrPdO$_3$ electrode – Towards a novel non-enzymatic glucose sensor, *Journal of Electroanalytical Chemistry*, Vol. 749, 2015, pp. 42-52.

[30]. L. Peng, S. Dong, H. Xie, G. Gu, Z. He, J. Lu, T. Huang, Sensitive simultaneous determination of diethylstilbestrol and bisphenol A based on Bi$_2$WO$_6$ nanoplates modified carbon paste electrode, *Journal of Electroanalytical Chemistry*, Vol. 726, 2014, pp. 15–20.

[31]. S. Shahrokhian, M. Ghalkhani, M. K. Amini, Application of carbon-paste electrode modified with iron phthalocyanine for voltammetric determination of epinephrine in the presence of ascorbic acid and uric acid, *Sensors and Actuators B*, Vol. 137, 2009, pp. 669–675.

[32]. S. Kitagawa, R. Kitaura, S. Noro, Functional porous coordination polymers, *Angewandte Chemie*, Vol. 43, 2004, pp. 2334–2375.

[33]. S. S. Y. Chui, S. M. F. Lo, J. P. H. Charmant, A. G. Orpen, I. D. Williams, A chemically functionalizable nanoporous material [Cu$_3$(TMA)$_2$(H2O)$_3$]n, *Science*, Vol. 283, 1999, pp. 1148–1150.

[34]. Y. Wang, Y. Wu, J. Xie, X. Hu, Metal–organic framework modified carbon paste electrode for lead sensor, *Sensors and Actuators B*, Vol. 177, 2013, pp. 1161–1166.

[35]. P. A. Gale, M. B. Hursthouse, M. E. Light, J. L. Sessler, C. N. Warriner, S. Zimmerman, Ferrocene-substituted calix[4]pyrrole: a new electrochemical sensor for anions involving CH···anion hydrogen bonds, *Tetrahedron Letters*, Vol. 42, 2001, pp. 6759-6762.

[36]. I. Szyman´ska, H. Radecka, J. Radecki, P. A. Gale, C. N. Warriner, Ferrocene-substituted calix[4]pyrrole modified carbon paste electrodes for anion detection in water, *Journal of Electroanalytical Chemistry*, Vol. 591, 2006, pp. 223–228.

[37]. N. F. Atta, A. Galal, F. M. Abdel-Gawad, E. F. Mohamed, Electrochemistry and Detection of Dobutamine at Gold Nanoparticles Cobalt-Phthalocyanine Modified Carbon Paste Electrode, *Journal of the Electrochemical Society*, Vol. 162, Issue 12, 2015, pp. B304-B311.

[38]. P. N. Mashazi, P. Westbroek, K. I. Ozoemena, T. Nyokong, Surface chemistry and electrocatalytic behaviour of tetra-carboxy substituted iron, cobalt and manganese phthalocyanine monolayers on gold electrode, *Electrochima Acta*, Vol. 53, 2007, pp. 1858–1869.

[39]. N. F. Atta, A. Galal, F. M. Abdel-Gawad, E. F. Mohamed, Electrochemical Morphine Sensor Based on Gold Nanoparticles Metalphthalocyanine Modified Carbon Paste Electrode, *Electroanalysis*, Vol. 27, 2015, pp. 415–428.

[40]. M. R. Ganjali, M. Hosseini, F. Basiripour, M. Javanbakht, O. R. Hashemi, M. F. Rastegar, M. Shamsipur, G. W. Buchanen, Novel coated-graphite

membrane sensor based on N, N_-dimethylcyanodiaza-18-crown-6 for the determination of ultra-trace amounts of lead, *Analytica Chimica Acta*, Vol. 464, 2002, pp. 181–186.

[41]. S. Cheraghi, M. Ali Taher, H. Fazelirad, Voltammetric sensing of thallium at a carbon paste electrode modified with a crown ether, *Microchimica Acta*, Vol. 180, 2013, pp 1157–1163.

[42]. K. S. Novoselov, A. K. Geim, S. V. Morozov, D. Jiang, Y. Zhang, S. V. Dubonos, I. V. Grigorieva, and A. A. Firsov, Electric Field Effect in Atomically Thin Carbon Films, *Science*, Vol. 306, 2004, pp. 666-669.

[43]. J. H. Chen, M. Ishigami, C. Jang, D. R. Hines, M. S. Fuhrer, E. D. Williams, Printed graphene circuits, *Advanced Materials*, Vol. 19, 2007, pp. 3623-3627.

[44]. J. Wintterling, M.-L. Bocquet, Graphene on metal surfaces, *Surface Science.*, Vol. 603, 2009, pp. 1841-1452.

[45]. C. J. Pool, On the applicability of the two-band model to describe transport across n–p junctions in bilayer graphene, *Solid State Communication*, Vol. 150, 2010, pp. 632-635.

[46]. L. Tang, Y. Wang, Y. Li, H. Feng, J. Lu, J. Li, Preparation, Structure, and Electrochemical Properties of Reduced Graphene Sheet Films, *Advanced Functional Materials*, Vol. 19, 2009, pp. 2782-2789.

[47]. L. Dong, R. R. S. Gari, Z. Li, M. M. Craig, S. Hou, Graphene-supported platinum and platinum–ruthenium nanoparticles with high electrocatalytic activity for methanol and ethanol oxidation, *Carbon*, Vol. 48, 2010, pp. 781-787.

[48]. Arxive website (http://arxiv.org/ftp/cond-mat/papers/0702/0702595.pdf)

[49]. J.-H. Chen, W. G. Cullen, C. Jang, M. S. Fuhrer, E. D. Williams, Defect scattering in graphene, *Physical Review Letter,* Vol. 102, issue 23, 2009, pp. 236805.

[50]. P. Lian, X. Zhu, S. Liang, Z. Li, W. Yang, H. Wang, Large reversible capacity of high quality graphene sheets as an anode material for lithium-ion batteries, *Electrochema Acta*, Vol. 55, 2010, pp. 3909-3914.

[51]. Y. Li, L. Tang, J. Li, On the applicability of the two-band model to describe transport across n–p junctions in bilayer graphene, *Electrochemistry Communications*, Vol. 11, 2009, pp. 846-849.

[52]. J. Wu, Y. Wang, D. Zhang, B. Hou, Studies on the electrochemical reduction of oxygen catalyzed by reduced graphene sheets in neutral media, *Journal of Power Sources*, Vol. 196, 2011, pp. 1141-1144.

[53]. S. Liu, J. Wang, J. Zeng, J. Ou, Z. Li, X. Liu, S. Yang, Green, Electrochemical synthesis of Pt/graphene sheet nanocomposite film and its electrocatalytic property, *Journal of Power Sources,* Vol. 195, 2010, pp. 4628-4633.

[54]. Y. B. Zhang, Y. W. Tan, H. L. Stormer, P. Kim, Experimental Observation of the Quantum Hall Effect and Berry's Phase in Graphene, *Nature,* Vol. 438, 2005, pp. 201-204.

[55]. K. S. Novoselov, D. Jiang, F. Schedin, T. J. Booth, V. V. Khotkevich, S. V. Morozov, A. K. Geim, Two-dimensional atomic crystals, *Proc Natl Acad Sci USA*, Vol. 102, 2005, pp. 10451-10453.

[56]. X. Lu, M. Yu, H. Huang and R. S. Ruoff, Tailoring graphite with the goal of achieving single sheets, *Nanotechnology*, Vol. 10, 1999, pp. 269-276.

[57]. F. S. Kim, Y. Zhao, H. Jang, S. Y. Lee, J. M. Kim, K. S. Kim, J. H. Ahn, P. Kim, J. Y. Choi, B. H. Hong, Large-scale pattern growth of graphene films for stretchable transparent electrodes, *Nature*, Vol. 457, 2009, pp. 706-710.

[58]. C. Berger, Z. Song, X. Li, X. Wu, N. Brown, C. Naud, D. Mayou, T. Li, J. Hass, A. N. Marchenkov, E. H. Conrad, P. N. First and W. A. de Heer, Electronic Confinement and Coherence in Patterned Epitaxial Graphene, *Science*, Vol. 312, 2006, pp. 1191-1196.

[59]. L. Jiao, L. Zhang, X. Wang, G. Diankov, H. Dai, Narrow graphene nanoribbons from carbon nanotubes, *Nature (London)*, Vol. 458, 2009, pp. 877-880.

[60]. A. Maiti, C. J. Brabec, C. Roland, J. Bernholc, Theory of carbon nanotube growth, *Physical Review B*, Vol. 52, 1995, pp. 14850-14858.

[61]. N. Li, Z. Y. Wang, K. K. Zhao, Z. J. Shi, Z. N. Gu, S. N. Xu, Large scale synthesis of N-doped multi-layered graphene sheets by simple arc-discharge method, *Carbon*, Vol. 48, 2010, pp. 255-259.

[62]. Y. Hu, J. Jin, P. Wu, H. Zhang, C. Cai, Graphene–gold nanostructure composites fabricated by electrodeposition and their electrocatalytic activity toward the oxygen reduction and glucose oxidation, *Electrochima Acta*, Vol. 56, Issue 1, 2010, pp. 491-500.

[63]. X. Wang, L. J. Zhi, N. Tsao, Z. Tomovic, J. L. Li, K. Mullen, Transparent Carbon Films as Electrodes in Organic Solar Cells, *Angewandte Chemie International Edition,* Vol. 47, 2008, pp. 2990-2992.

[64]. J. Hass, W. A. de Heer, E. H. Conrad, The growth and morphology of epitaxial multilayer graphene*, Journal Physics: Condensed Matter*, Vol. 20, 2008, pp. 323202.

[65]. M. D. Stoller, S. J. Park, Y. W. Zhu, J. H. An, R. S. Ruoff, Graphene-Based Ultracapacitors, *Nano Letter*, Vol. 8, 2008, pp. 3498-3502.

[66]. E. Yoo, J. Kim, E. Hosono, H. Zhou, T. Kudo, I. Honma, arge Reversible Li Storage of Graphene Nanosheet Families for Use in Rechargeable Lithium Ion Batteries, *Nano Letter*, Vol. 8, 2008, pp. 2277-2279.

[67]. D. H. Wang, D. W. Choi, J. Li, Z. G. Yang, Z. M. Nie, R. Kou, D. H. Hu, C. M. Wang, L. V. Saraf, J. G. Zhang, I. A. Aksay, J. Liu, Self-Assembled TiO_2–Graphene Hybrid Nanostructures for Enhanced Li-Ion Insertion, *ACS Nano,* Vol. 3, 2009, pp. 907-914.

[68]. B. Seger, P. V. Kamat, Electrocatalytically Active Graphene-Platinum Nanocomposites. Role of 2-D Carbon Support in PEM Fuel Cells, *Journal of Physical Chemistry C*, Vol. 113, 2009, pp. 7990-7995.

[69]. R. Kou, Y. Shao, D. H. Wang, M. H. Engelhard, J. H. Kwak, J. Wang, V. V. Viswanathan, C. M. Wang, Y. H. Lin, Y. Wang, I. A. Aksay, J. Liu, Enhanced activity and stability of Pt catalysts on functionalized graphene sheets for electrocatalytic oxygen reduction, *Electrochemistry Communications*, Vol. 11, 2009, pp. 954-957.

[70]. Y. C. Si, E. T. Samulski, Exfoliated Graphene Separated by Platinum Nanoparticles, *Chemistry of Materials*, Vol. 20, 2008, pp. 6792-6797.

[71]. Y. M. Li, L. H. Tang, J. H. Li, Preparation and electrochemical performance for methanol oxidation of Pt/graphene nanocomposites, *Electrochemistry Communications*, Vol. 11, 2009, pp. 846-849.

[72]. Y. Wang, J. Lu, L. H. Tang, H. X. Chang, J. H. Li, Graphene Oxide Amplified Electrogenerated Chemiluminescence of Quantum Dots and its Selective Sensing for Glutathione from Thiol-Containing Compounds, *Analytical Chemistry*, Vol. 81, 2009, pp. 9710-9715.

[73]. Y. Jiang, X. Zhang, C. Shan, S. Hua, Q. Zhang, X. Bai, L. Dan, L. Niu, Functionalization of graphene with electrodeposited Prussian blue towards amperometric sensing application, *Talanta*, Vol. 85, 2011, pp. 76–81.

[74]. S. Liu, J. Tian, L. Wang, X, Sun, Electrochemical detection of dopamine in the presence of ascorbic acid using graphene modified electrodes, *Journal of Nanoparticles Research*, Vol. 13, Issue 10, 2011, pp. 4539-4548.

[75]. F. Li, J. Chai, H. Yang, D. Han, L. Niu, Synthesis of Pt/ionic liquid/graphene nanocomposite and its simultaneous determination of ascorbic acid and dopamine, *Talanta,* Vol. 81, Issue 8, 2010, pp. 1063-1068.

[76]. L. Luo, L. Zhu, Z. Wang, Nonenzymatic amperometric determination of glucose by CuO nanocubes–graphene nanocomposite modified electrode, *Bioelectrochemistry*, Vol. 88, pp. 156-163.

[77]. J. B. Raoof, A. Kiani, R. Ojani, R. Valiollahi, S. Rashid-Nadimi, Simultaneous voltammetric determination of ascorbic acid and dopamine at the surface of electrodes modified with self-assembled gold nanoparticle films, *Journal of Solid State Electrochemistry*, Vol. 14, 2010, pp. 1171–1176.

[78]. H. K. Hassan, N. F. Atta, A. Galal, Rapid and facile electrochemical detection of morphine on graphene-palladium hybrid modified glassy carbon electrode, *Anal Bioanal Chem.*, Vol. 406, Issue 27, 2014, pp. 6933-6942.

[79]. K. Liu, J. Zhang, G. Yang, C. Wang, J.-J. Zhu, Direct electrochemistry and electrocatalysis of hemoglobin based on poly(diallyldimethylammonium chloride) functionalized graphene sheets/room temperature ionic liquid composite film, *Electrochemistry Communications*, Vol. 12 2010, pp. 402–405.

[80]. D. X. Han, T. T. Han, C. S. Shan, A. Ivaska, L. Niu, Simultaneous Determination of Ascorbic Acid, Dopamine and Uric Acid with Chitosan-

Graphene Modified Electrode, *Electroanalysis*, Vol. 22, 2010, pp. 2001–2008.

[81]. R. M. D. Carvalho, R. S. Freire, S. Rath, L. T. Kubota, Effects of EDTA on signal stability during electrochemical detection of acetaminophen, *J. Pharmaceutical and Biomedical Analalysis,* Vol. 34, 2004, pp. 871-878.

[82]. Y. Jiang, Q. Zhanga, F. Li, L. Niu, Glucose oxidase and graphene bionanocomposite bridged by ionic liquid unit for glucose biosensing application, *Sensors and Actuators B*, Vol. 161, 2012, pp. 728– 733.

[83]. S. Hu, Y. Wang, X. Wang, L. Xu, J. Xiang, W. Sun, Electrochemical detection of hydroquinone with a gold nanoparticle and graphene modified carbon ionic liquid electrode, *Sensors and Actuators B*, Vol. 168, 2012, pp. 27– 33.

[84]. A. Sinitskii, A. Dimiev, D. A. Corley, A. A. Fursina, D. V. Kosynkin, J. M. Tour, Kinetics of Diazonium Functionalization of Chemically Converted Graphene Nanoribbons, *ACS Nano*, Vol. 4, 2010, pp. 1949-1954.

[85]. Sofo, J. O., Chaudhari, A. S., Barber, G. D., Graphane: A two-dimensional hydrocarbon, *Physical Review B*, Vol. 75, 2007, p. 153401.

[86]. D. C. Elias, R. R. Nair, T. M. G. Mohiuddin, S. V. Morozov, P. Blake, M. P. Halsall, A. C. Ferrari, D. W. Boukhvalov, M. I. Katsneltson, M. I., Geim, K. S. Novoselov, Spectral Modifications of Graphene Using Molecular Dynamics Simulations, *Science*, Vol. 323, 2009, pp. 610-613.

[87]. Chandrachud, P., Pujari, B. S., Haldar, S., Sanyal, B., Kanhere, D. G., A systematic study of electronic structure from graphene to graphene, *J. Phys.: Condens. Matter,* 2010, 22, No. 465502.

[88]. M. Z. S. Flores, P. A. S. Autreto, S. B. Legolas, D. S. Galvao, Graphene to graphane: a theoretical study, *Nanotechnology*, Vol. 20, 2009, p. 465704.

[89]. R. Zboril, F. Karlicky, A. B. Bourlinos, B. Athanasios, T. A. Steriotis, A. A. K. Stubos, V. Georgakilas, K. Safarova, D. Jancik, C. Trapalis, M. Otyepka, Graphene Fluoride: A Stable Stoichiometric Graphene Derivative and its Chemical Conversion to Graphene, *Small,* Vol. 6, 2010, pp. 2885-2891.

[90]. Zhou, L., Wang, Q., Sun, Q., Chen, X., Kawazoe, Y., Jena, P., Ferromagnetism in Semihydrogenated Graphene Sheet, *Nano Lett.*, 9, 2009, pp. 3867-3870.

[91]. D. Elias, R. Nair, T. Mohiuddin, S. Morozov, P. Blake, M. Halsall, A. Ferrari, D. Boukhvalov, M. Katsnelson, A. Geim, K. Novoselov, Control of Graphene's Properties by Reversible Hydrogenation: Evidence for Graphane, *Science*, Vol. 323, 2009, pp. 610-613.

[92]. J. T. Robinson, J. S. Burgess, C. E. Junkermeier, S. C. Badescu, T. L. Reinecke, F. K. Perkins, M. K. Zalalutdniov, J. W. Baldwin, J. C. Culbertson, P. E. Sheehan, E. S. Snow, Properties of Fluorinated Graphene Films, *Nano Letter*, Vol. 10, 2010, pp. 3001-3005.

[93]. F. Withers, M. Dubois, A. K. Savchenko, Electron properties of fluorinated single-layer graphene transistors, *Physical Review B*, Vol. 82, 2010, pp. 073403.

[94]. O. Jankovský, A. Libánská, D. Bouša, D. Sedmidubský, S. Matějkov, Z. Sofer, Partially Hydrogenated Graphene Materials Exhibit High Electrocatalytic Activities Related to Unintentional Doping with Metallic Impurities, *Chemistry - A European Journal*, Vol. 22, 2016, pp. 8627 – 8634.

[95]. M. S. Islam, T. Hussain, G. S. Rao, P. Panigrahi, Rajeev Ahuja Augmenting the sensing aptitude of hydrogenated graphene by crafting with defects and dopants, *Sensors and Actuators B: Chemical*, Vol. 228, 2016, pp. 317–321.

[96]. H. Zhang, L. Fan, H. Dong, P. Zhang, K. Nie, J. Zhong, Y. Li, J. Guo, and X. Sun, Spectroscopic Investigation of Plasma-Fluorinated Monolayer Graphene and Application for Gas Sensing, *ACS Applied Materials & Interfaces*, Vol. 8, Issue 13, 2016, pp. 8652–8661.

[97]. L. Yan, M. Lin, C. Zeng, Z. Chen, X. Zhao, A. Wu, Y. Wang, S. Zhang, J. Qu, L. Dai, M. Guo, Y. Liu, Electroactive and Biocompatible Hydroxyl-functionalized Graphene by Ball Milling, *Journal of Material Chemistry*, Vol. 22, 2012, pp. 8367-8371.

[98]. J. Sun, Y. Deng, J. Li, G. Wang, P. He, S. Tian, X. Bu, Z. Di, S. Yang, Guqiao Ding, and Xiaoming Xie, A New Graphene Derivative: Hydroxylated Graphene with Excellent Biocompatibility, *ACS Applied Materials & Interfaces*, Vol. 8, Issue 16, 2016, pp. 10226–10233.

[99]. J. Wu, K. Tao, J. Zhang, Y. Guo, J. Miao, L. K. Norford, Chemically functionalized 3D graphene hydrogel for high performance gas Sensing, *Journal of Materials Chemistry A*, Vol. 4, 2016, pp. 8130-8140.

[100]. D. V. Kosynkin, A. L. Higginbotham, A. Sinitskii, J. R. Lomeda, A. Dimiev, B. K. Price, J. M. Tour, Longitudinal unzipping of carbon nanotubes to form graphene nanoribbons, *Nature*, Vol. 458, 2009, pp. 872-876.

[101]. S. Niyogi, E. Bekyarova, M. E. Itkis, H. Zhang, K. Shepperd, J. Hicks, M. Sprinkle, C. Berger, C. Ning Lau, W. A. de Heer, E. H. Conrad, R. C. Haddon, Spectroscopy of Covalently Functionalized Graphene, *Nano Letter*, Vol. 10, 2010, pp. 4061.

[102]. Z. Jin, J. R. Lomeda, B. K. Price, W. Lu, Y. Zhu, J. M. Tour, Mechanically Assisted Exfoliation and Functionalization of Thermally Converted Graphene Sheets, *Chemistry of Materials*, 2009, 21, pp. 3045-3047.

[103]. J. R. Lomeda, C. D. Doyle, D. V. Kosynkin, W. F. Hwang, J. M. Tour, Diazonium Functionalization of Surfactant-Wrapped Chemically Converted Graphene Sheets, *Journal of American Chemical Society*, Vol. 130, 2008, 16201-16206.

[104]. E. Bekyarova, M. E. Itkis, P. Ramesh, C. Berger, M. Sprinkle, W. A. de Heer, R. C. Haddon, Chemical Modification of Epitaxial

Graphene: Spontaneous Grafting of Aryl Groups, *Journal of American Chemical Society,* Vol. 131, 2009, pp. 1336-1337.

[105]. M. Z. Hossain, M. A. Walsh, M. C Hersam, Scanning Tunneling Microscopy, Spectroscopy, and Nanolithography of Epitaxial Graphene Chemically Modified with Aryl Moieties, *Journal of American Chemical Society,* Vol. 132, 2010, pp. 15399-15403.

[106]. M. Fang, K. Wang, H. Lu, Y. Yang, S. Nutt, Covalent polymer functionalization of graphene nanosheets and mechanical properties of composites, *Journal of Material Chemistry,* Vol. 19, 2009, pp. 7098–7105.

[107]. H. Liu, S. Ryu, Z. Chen, M. L. Steigerwald, C. Nuckolls, L. E. Brus, Photochemical Reactivity of Graphene, *Journal of American Chemical Society,* Vol. 131, 2009, pp. 17099-17101.

[108]. V. Georgakilas, M. Otyepka, A. B. Bourlinos, V. Chandra, N. Kim, K. C. Kemp, P. Hobza, R. Zboril, K. S. Kim, Functionalization of Graphene: Covalent and Non-Covalent Approaches, Derivatives and Applications, *Chemical Review,* Vol. 112, 2012, pp. 6156–6214.

[109]. V. Georgakilas, A. B. Bourlinos, D. Gournis, T. Tsoufis, C. Trapalis, A. M. Alonso, M. Prato, Multipurpose Organically Modified Carbon Nanotubes: From Functionalization to Nanotube Composites, *Journal of American Chemical Society,* Vol. 130, 2008, pp. 8733-8740.

[110]. V. Georgakilas, D. M. Guldi, R. Signorini, R. Bozio, M. Prato, Organic Functionalization and Optical Properties of Carbon Onions, *Journal of American Chemical Society,* Vol. 125, 2003, PP. 14268-14269.

[111]. V. Georgakilas, A. B. Bourlinos, R. Zboril, T. A. Steriotis, P. Dallas, A. K. Stubos, C. Trapalis, Organic functionalisation of graphenes, *Chemical Communications,* Vol. 46, 2010, pp. 1766-1768.

[112]. X. Zhang, L. Hou, A. Cnossen, A. C. Coleman, O. Ivashenko, P. Rudolf, B. J. van Wees, W. R. Browne, B. L. Feringa, One-Pot Functionalization of Graphene with Porphyrin through Cycloaddition Reactions, *Chemistry - A European Journal,* Vol. 17, 2011, pp. 8957-8964.

[113]. L. H. Liu, M. M. Lerner, M. Yan, Derivitization of pristine graphene with well-defined chemical functionalities, *Nano Letters,* Vol. 10, 2010, pp. 3754-3756.

[114]. S. Vadukumpully, J. Gupta, Y. Zhang, C. Q. Xu, S. Valiyaveettil, Functionalization of surfactant wrapped graphene nanosheets with alkylazides for enhanced dispersibility, *Nanoscale,* Vol. 3, 2011, pp. 303-308.

[115]. H. He, C. Gao, General Approach to Individually Dispersed, Highly Soluble, and Conductive Graphene Nanosheets Functionalized by Nitrene Chemistry, *Chemistry of Materials,* Vol. 22, 2010, pp. 5054-5064.

[116]. Y. Liu, J. Zhou, X. Zhang, Z. Liu, X. Wan, J. Tian, T. Wang, Y. Chen, Synthesis, characterization and optical limiting property of covalently

oligothiophene-functionalized graphene material, *Carbon*, Vol. 47, 2009, pp. 3113-3121.

[117]. D. Yu, Y. Yang, M. Durstock, J. B. Baek, L. Dai, Soluble P3HT-Grafted Graphene for Efficient Bilayer–Heterojunction Photovoltaic Devices, *ACS Nano*, Vol. 4, 2010, pp. 5633-5640.

[118]. N. Karousis, A. S. D. Sandanayaka, T. Hasobe, S. P. Economopoulos, E. Sarantopoulou, N. Tagmatarchis, Graphene oxide with covalently linked porphyrin antennae: Synthesis, characterization and photophysical properties, *Journal of Materials Chemistry*, Vol. 21, 2011, pp. 109-114.

[119]. X. Zhang, Y. Feng, D. Huang, Y. Li, W. Feng, Investigation of optical modulated conductance effects based on a graphene oxide–azobenzene hybrid, *Carbon*, Vol. 48, 2010, pp. 3236-3241.

[120]. Y. Xu, Z. Liu, X. Zhang, Y. Wang, J. Tian, Y. Huang, Y. Ma, X. Zhang, Y. Chen, A Graphene Hybrid Material Covalently Functionalized with Porphyrin: Synthesis and Optical Limiting Property, *Advanced Materials*, Vol. 21, 2009, pp. 1275-1279.

[121]. K. P. Pramoda, H. Hussain, H. M. Koh, H. R. Tan, C. B. He, Covalent bonded polymer–graphene nanocomposites, *Journal of Polymer Science, Part A: Polymer Chemistry*, Vol. 48, 2010, pp. 4262-4267.

[122]. H. Yang, C. Shan, F. Li, D. Han, Q. Zhang, L. Niu, Covalent functionalization of polydisperse chemically-converted graphene sheets with amine-terminated ionic liquid, *Chemical Communications*, Vol. 26, 2009, pp. 3880-3882.

[123]. N. Karousis, S. P. Economopoulos, E. Sarantopoulou, N. Tagmatarchis, Porphyrin counter anion in imidazolium-modified graphene-oxide, *Carbon*, Vol. 48, 2010, pp. 854-860.

[124]. D. Kumar, S. Rani, Synthesis of Amide Functionalized Graphene Oxide for Humidity Sensing Application (Amide Functionalized Graphene Oxide for Humidity Sensing), in *Proceedings of the 6th International Conference on Sensor Device Technologies and Applications*, 2015, pp. 142-145.

[125]. K.-J. Huang, D.-J. Niu, X. Liu, Z-W. Wu, Y. Fan, Y.-F. Chang, Y.-Y. Wu, Direct electrochemistry of catalase at amine-functionalized graphene/gold nanoparticles composite film for hydrogen peroxide sensor *Electrochimica Acta*, Vol. 56, 2011, pp. 2947–2953.

[126]. K. Hasan, M. H. Asif, O. Nur, M. Willander, Needle-Type Glucose Sensor Based on Functionalized Graphene, *Journal of Biosensors and Bioelectronics*, Vol. 3, Issue 1, 2012, pp. 2155-6210.

[127]. G. Jiang, M. Goledzinowski, F. J. E. Comeau, H. Zarrin, G. Lui, J. Lenos, A. Veileux, G. Liu, J. Zhang, S. Hemmati, J. Qiao, Z. Chen, Free-Standing Functionalized Graphene Oxide Solid Electrolytes in Electrochemical Gas Sensors, *Advanced Functional Materials*, Vol. 26, 2016, pp. 1729–1736.

[128]. M. A. Bissett, M. Tsuji, H. Ago, Mechanical Strain of Chemically Functionalized Chemical Vapor Deposition Grown Graphene, *Journal of Physical Chemistry C,* Vol. 117, 2013, pp. 3152–3159.

[129]. H. Zhang, E. Bekyarova, J.-W. Huang, Z. Zhao, W. Bao, F. Wang, R. C. Haddon, C. N. Lau, Aryl Functionalization as a Route to Band Gap Engineering in Single Layer Graphene Devices, *Nano Letters,* Vol. 11, 2011, pp. 4047–4051.

[130]. Q. H. Wang, Z. Jin, K. K. Kim, A. J. Hilmer, G. L. C. Paulus, C.-J. Shih, M.-H. Ham, J.-D. Sanchez-Yamagishi, K. Watanabe, T. Taniguchi, J. Kong, P. Jarillo-Herrero, M. S. Strano, Understanding and controlling the substrate effect on graphene electron-transfer chemistry via reactivity imprint lithography, *Nature Chemistry,* Vol. 7, 2012, pp. 724–732.

[131]. F. Guinea, M. I. Katsnelson, A. K. Geim, Energy gaps and a zero-field quantum Hall effect in graphene by strain engineering, *Nature Physics,* Vol. 6, 2010, pp. 30–33.

[132]. K.-J. Huang, D.-J. Niu, J.-Y. Sun, C.-H. Han, Z.-W. Wu, Y.-L. Li, X.-Q. Xiong, Novel electrochemical sensor based on functionalized graphene for simultaneous determination of adenine and guanine in DNA, *Colloids and Surfaces B: Biointerfaces,* Vol. 82, 2011, pp. 543–549.

[133]. Y. Hu, F. Li, X. Bai, D. Li, S. Hua, K. Wanga, L. Niu, Label-free electrochemical impedance sensing of DNA hybridization based on functionalized graphene sheets, *Chemical Communications,* Vol. 47, 2011, pp. 1743–1745.

[134]. F. Li, H. Yang, C. Shan, Q. Zhang, D. Han, A. Ivaska, L. Niu, The synthesis of perylene-coated graphene sheets decorated with Au nanoparticles and its electrocatalysis toward oxygen reduction, *Journal of Materials Chemistry,* Vol. 19, 2009, pp. 4022-4025.

[135]. X. R. Wang, X. L. Li, L. Zhang, Y. Yoon, P. K. Weber, H. L. Wang, J. Guo, H. J. Dai, N-Doping of Graphene Through Electrothermal Reactions with Ammonia, *Science,* Vol. 324, 2009, pp. 768-771.

[136]. X. Li, H. Wang, J. T. Robinson, H. Sanchez, G. Diankov, H. Dai, Simultaneous Nitrogen Doping and Reduction of Graphene Oxide, *Journal of American Chemical Society,* Vol. 131, 2009, pp. 15939-15944.

[137]. D. Long, W. Li, L. Ling, J. Miyawaki, I. Mochida, S. H. Yoon, Preparation of Nitrogen-Doped Graphene Sheets by a Combined Chemical and Hydrothermal Reduction of Graphene Oxide, *Langmuir,* Vol. 26, 2010, pp. 16096-16102.

[138]. D. Wei, Y. Liu, Y. Wang, H. Zhang, L. Huang, G. Yu, Synthesis of N-Doped Graphene by Chemical Vapor Deposition and Its Electrical Properties, *Nano Letters,* Vol. 9, 2009, pp. 1752-1758.

[139]. L. S Panchokarla, K. S. Subrahmanyam, K. S. Saha, A. Govindaraj, H. R. Krishnamurthy, U. V. Waghmare, C. N. R. Rao, Synthesis,

Structure, and Properties of Boron- and Nitrogen-Doped Graphene, *Advanced Materials*, Vol. 21, 2009, pp. 4726-4730.

[140]. L. Zhao, R. He, K. T. Rim, T. Schiros, K. S. Kim, H. Zhou, C. Gutiérrez, S. P. Chockalingam, C. J. Arguello, L. Pálová, D. Nordlund, M. S. Hybertsen, D. R. Reichman, T. F. Heinz, P. Kim, A. Pinczuk, G. W. Flynn, A. N. Pasupathy, Visualizing individual nitrogen dopants in monolayer graphene, *Science,* Vol. 333, 2011, pp. 999.

[141]. Z.-H. Shenga, X.-Q. Zheng, J.-Y. Xu, W.-J. Bao, F.-B. Wang, X.-H. Xia, Electrochemical sensor based on nitrogen doped graphene: Simultaneous determination of ascorbic acid, dopamine and uric acid, *Biosensors and Bioelectronics,* Vol. 34, 2012, pp. 125–131.

[142]. X. Feng, Y. Zhang, J. Zhou, Y. Li, S. Chen, L. Zhang, Y. Ma, L. Wang, X. Yan, Three-dimensional nitrogen-doped graphene as an ultrasensitive electrochemical sensor for the detection of dopamine, Nanoscale, Vol. 7, Issue 6, 2015, pp. 2427-2432.

[143]. H. Fan, Y. Li, D. wu, Q. Wei, Electrochemical bisphenol A sensor based on N-doped graphene sheets, *Analytica Chimica Acta,* Vol. 711, 2012, pp. 24-28 ·.

[144]. S. Luo, Y. Chen, A. Xie, Y. Kong, B. Wang, C. Yao, Nitrogen Doped Graphene Supported Ag Nanoparticles as Electrocatalysts for Oxidation of Glucose, *Journal of Electrochemical Society,* Vol. 3, 2014, pp. B20-B22.

[145]. R. Lv, G. Chend, Q. Lie, A. McCreary, A. Botello-Méndezf, S. V. Morozov, L. Liangh, X. Declerck, N. Perea-López, D. A. Culleni, S. Feng, A. L. Elías, R. Cruz-Silvaj, K. Fujisaw, M. Endoj, F. Kanga, J.-C. Charlier, V. Meunierh, M. Pank, A. R. Harutyuny, K. S. Novoselov, M. Terrones, Ultrasensitive gas detection of large-area boron-doped graphene, *PNAS,* Vol. 112, 2015, pp. 14527–14532.

[146]. H. Gao, Z. Liu, L. Song, W. Guo, W. Gao, L. Ci, A. Rao, W. Quan, R. Vajtai, P. M. Ajayan, Synthesis of S-doped graphene by liquid precursor, *Nanotechnology,* Vol. 23, 2012, pp. 275605-275613.

[147]. E. Z. Galakhov, A. Moewes, S. Moehlecke, Y. Kopelevich, Interlayer conduction band states in graphite-sulfur composites, *Physical Review B*, Vol. 66, 2002, pp. 193402.

[148]. P. A. Denis, Band gap opening of monolayer and bilayer graphene doped with aluminium, silicon, phosphorus, and sulfur, *Chemical Physics Letter*, Vol. 492, 2010, pp. 251-257.

[149]. M. AoZ, J. Yang, S. Li, Q. Jiang, Enhancement of CO detection in Al doped graphene, *Chemical Physics Letters*, Vol. 461, 2008, pp. 276-279.

[150]. C. Liang, Y. L. Wang, T. Li, Sulfur-doping in graphene and its high selectivity gas sensing in NO_2, in *Proceedings of the 18th International Conference on Solid-State Sensors, Actuators and Microsystems (TRANSDUCERS' 15),* Anchorage, Alaska, USA, June 21-25, 2015, W2P.027.

[151]. Y. Liu, Y. Ma, Y. Jin, G. Chen, X. Zhang, Microwave-assisted solvothermal synthesis of sulfur-doped graphene for electrochemical

Sensing, *Journal of Electroanalytical Chemistry*, Vol. 739, 2015, pp. 172–177 S.

[152]. Yang, L. Zhi, K. Tang, X. Feng, J. Maier, K. Müllen, Efficient Synthesis of Heteroatom (N or S)-Doped Graphene Based on Ultrathin Graphene Oxide-Porous Silica Sheets for Oxygen Reduction Reactions, *Advanced Functional Materials*, Vol. 22, 2012, pp. 3634-3640.

[153]. H. L. Poh, P. S imek, Z. k. Sofer, M. Pumera, Sulfur-Doped Graphene via Thermal Exfoliation of Graphite Oxide in H_2S, SO_2, or CS_2 Gas, *ACS Nano*, Vol. 7, 2013, pp. 5262-5272.

[154]. D. Higgins, M. A. Hoque, M. H. Seo, R. Wang, F. Hassan, J. Y. Choi, M. Pritzker, A. Yu, J. Zhang, Z. Chen, Advanced Functional Materials, Vol. 27, 2014, pp. 4325-4336.

[155]. F. Niu, L.-M. Tao, Y.-C. Deng, Q.-H. Wang, W.-G. Song, Phosphorus doped graphene nanosheets for room temperature NH_3 sensing, *New Journal of Chemistry*, Vol. 38, 2014, pp. 2269—2272.

[156]. V. ThirumalA. Pandurangan, R. Jayavel, K. S. Venkatesh, N. S. Palani R. Ragavan, R. Ilangovan, Single pot electrochemical synthesis of functionalized and phosphorus doped graphene nanosheets for supercapacitor applications, *Journal of Materials Science: Materials in Electronics*, Vol. 26, 2015, pp. 6319–6328.

[157]. J. Liang, Y. Jiao, M. Jaroniec, S. Z. Qiao, Sulfur and nitrogen dual-doped mesoporous graphene electrocatalyst for oxygen reduction with synergistically enhanced performance, *Angewandte Chemie International Edition*, Vol. 51, 2012, pp. 11496-11500.

[158]. G. Chen, Y. Liu, Y. Liu, Y. Tian, X. Zhang, Nitrogen and sulfur dual-doped graphene for glucose biosensor application, *Journal of Electroanalytical Chemistry*, Vol. 738, 2015, pp. 100-107.

[159]. G. H. Yang, Y. H. Zhou, J. J. Wu, J. T. Cao, L. L. Li, H. Y. Liu, J. J. Zhu, Microwave-assisted synthesis of nitrogen and boron co-doped graphene and its application for enhanced electrochemical detection of hydrogen peroxide, *RSC Advances*, Vol. 3, 2013, pp. 22597-22604.

[160]. P. Guo, F. Xiao, Q. Liu, H. Liu, Y. Guo, J. R. Gong, S. Wang, Y. Liu, One-Pot Microbial Method to Synthesize Dual-Doped Graphene and Its Use as High-Performance Electrocatalyst, *Scientific Report*, Vol. 3, 2013, pp. 3499-3504.

[161]. I. Choudhuri, N. Patra, A. Mahata, R. Ahuja, B. Pathak, B-N@Graphene: Highly Sensitive and Selective Gas Sensor, *Journal of Physical Chemistry C*, Vol. 119, 2015, pp. 24827–24836.

[162]. M. Chicharro, A. Zapardiel, E. Bermejo, M. Moreno, E. Madrid, Electrocatalytic amperometric determination of amitrole using a cobalt-phthalocyanine-modified carbon paste electrode, *Analytical and Bioanalytical Chemistry*, Vol. 373, 2002, pp. 277-283.

[163]. J. Guo, Y. Chai, R. Yuan, Z. Song, Z. Zou, Lead (II) carbon paste electrode based on derivatized multi-walled carbon nanotubes:

Application to lead content determination in environmental samples, *Sensors and Actuators B*, Vol. 155, 2011, pp. 639–645.

[164]. X. Liu, L. Luo, Y. Ding, D. Ye, Poly-glutamic acid modified carbon nanotube-doped carbon paste electrode for sensitive detection of L-tryptophan, *Bioelectrochemistry*, Vol. 82, 2011, pp. 38–45.

[165]. Hassan Karimi-Maleh, Pourya Biparva, Mehdi Hatami, A novelmodified carbon paste electrode based on NiO/CNTs nanocomposite and (9, 10-dihydro-9,10-ethano anthracene-11,12-dicarboximido)-4-ethyl benzene-1,2-diol as a mediator for simultaneous determination of cysteamine, nicotin amide adenine dinucleotide and folic acid, *Biosensors and Bioelectronics*, Vol. 48, 2013, pp. 270–275.

[166]. S. Shahrokhian, S. Rastgar, M. K. Amini, M. Adeli, Fabrication of a modified electrode based on Fe$_3$O$_4$NPs/MWCNT nanocomposite: application to simultaneous determination of guanine and adenine in DNA, *Bioelectrochemistry* Vol. 86, 2012, pp. 78–86.

[167]. H. Yin, Y. Zhou, Q. Ma, Sh. Ai, Q. Chen, L. Zhu, Electrocatalytic oxidation behavior of guanosine at graphene, chitosan and Fe3O4 nanoparticles modified glassy carbon electrode and its determination, *Talanta*, Vol. 82, 2010, pp. 1193–1199.

[168]. H. Bagheri, A. Afkhami, Y. Panahi, H. Khoshsafar, A. Shirzadmehr, Facile stripping voltammetric determination of haloperidol using a high performance magnetite/carbon nanotube paste electrode in pharmaceutical and biological samples, *Materials Science and Engineering C*, Vol. 37, 2014, pp. 264–270.

[169]. A. Benvidi, P. Dehghan, A. D. Firouzabadi, H. Emtiazi, H. R. Zare, M. Mazloum-Ardakani, Construction of a nanocomposite sensor by the modification of a carbon-paste electrode with reduced graphene oxide and a hydroquinone derivative: simultaneous determination of glutathione and penicillamine, *Analytical Methods*, 7, 2015, pp. 5538-5544.

[170]. S. Tajik, M. A. Taher, H. Beitollahi, The first electrochemical sensor for determination of mangiferin based on an ionic liquid–graphene nanosheets paste electrode, *Ionics,* Vol. 20, 2014, pp. 1155–1161.

[171]. Sanchez G. M., Re L., Giuliani A., Nunez-Selles A. J., Davison G. P., Leon Fernandez O. S., Protective effects of Mangifera indica L. extract, mangiferin, and selected antioxidants against TPA-induced biomolecules oxidation and peritoneal macrophage activation in mice, *Pharm Res*, 42, 2000, pp. 565-573.

[172]. S. Guha, S. Ghosal, U. Chattopadhyay, Antitumor, Immunomodulatory and Anti-HIV Effect of Mangiferin, a Naturally Occurring Glucosylxanthone, *Chemotherapy*, Vol. 42, 1996, pp. 443-451.

[173]. D. Garcia, M. Escalante, R. Delgado, F. M. Ubeira, J. Leiro, You have full Anthelminthic and antiallergic activities of *Mangifera indica* L. stem bark components Vimang and mangiferin, *Phytotherapy Research,* Vol. 17, 2003, pp. 1203-1208.

[174]. Garcia D, Escalante M, Delgado R, Ubeira FM, J. Leiro, Anthelminthic and anti-allergic activities of *Mangifera indica* L. stem bark components Vimang and mangiferin, *Phytotherapy Research*, 17, 2003, pp. 1203-1208.

[175]. H. Z. Lee, W. C. Lin, F. T. Yeh, C. N. Lin, C. H. Wu, Decreased protein kinase C activation mediates inhibitory effect of norathyriol on serotonin-mediated endothelial permeability, *European Journal Pharmacology*, Vol. 353, 1998, pp. 303-313.

[176]. T. Miura, N. Iwamoto, M. Kato, H. Ichiki, M. Kubo, Y. Komatsu, Antidiabetic activity of the rhizoma of Anemarrhena asphodeloides and active components, mangiferin and its glucoside, *Biological and Pharmaceutical Bulletin*, Vol. 24, 2001, pp. 1009-1011.

[177]. M. Yoshikawa, K. Ninomiya, H. Shimoda, N. Nishida, H. Matsuda, Hepatoprotective and antioxidative properties of Salacia reticulata: preventive effects of phenolic constituents on CCl_4-induced liver injury in mice, *Biological and Pharmaceutical Bulletin*, Vol. 25, 2002, pp. 72-76.

[178]. X. Yuan, Y. Chai, R. Yuan, Q. Zhao, C. Yang, Functionalized graphene oxide-based carbon paste electrode for potentiometric detection of copper ion (II), *Analytical Methods,* Vol. 4, 2012, pp. 3332–3337.

[179]. J. Shi, X. Li, H. Cheng, Z. Liu, L. Zhao, T. Yang, Z. Dai, Z. Cheng, E. Shi, L. Yang, Z. Zhang, A. Cao, H. Zhu, Y. Fang, Graphene Reinforced Carbon Nanotube Networks for Wearable Strain Sensors, *Advanced Functional Materials*, Vol. 26, 2016, pp. 2078–2084.

Index

A

Analytical characterization of drug-CD complexes, *72*
  Capillary electrophoresis, *77, 78*
  Circular dichroism spectroscopy, *73*
  Electrical conductivity, *76*
  Electroanalytical techniques, *75*
  Electron spin resonance, *75*
  Fluorescence spectroscopy, *73*
  HPLC, *77*
  Isothermal titration calorimetry, *79*
  NMR spectroscopy, *74*
  Polarimetry, *78*
  Polarography and voltammetry, *75*
  Potentiometry, *75, 76*
  Separation techniques, *77*
  Spectroscopic techniques, *72*
  UV-vis spectroscopy, *72*
Applications of cyclodextrin
  Cholesterol free products, *67*
  Dopamine (DA), *69, 70*
  Environmental protection, *71*
  Graphene, *70, 71*
  HPLC columns, *72*
  Ionic liquid crystal (ILC), *70*
  Paracetamol, *69*
  Pharmaceutical field, *68, 69*
  Supramolecular chemistry, *71*
Applications of ferrocene, *57*
  Cobaltocene (CC), *60*
  DNA sensor, *60*
  Drug analysis, *58*
  Ferrocene (FC2), *60*
  Ferrocene carboxylic acid (FC1), *60*
  Fuel additives, *57*
  Ion-selective electrodes, *61*
  Layer-by-layer deposition, *59*
  Ligands, *58*
  Materials chemistry, *58*
  Morphine, *60*
  Voltammetric measurements, *58*
Applications of perovskites
  Carbon paste electrode (CPE), *184, 186*
  Core-shell, *190*
  Dopamine (DA), *184, 186, 188*
  Dopant, *183*
  Gas sensor, *179, 180*

255

Glucose sensor, *180, 185*
Gold nanoparticles (Aunano), *181*
Graphite electrode, *186*
H_2O_2 sensor, *180, 185*
Ionic liquid crystal (ILC), *190, 191*
Ketotifen, *189*
$LaFeO_3$, *188*
L-dopa, *186*
Metoclopramide, *190, 191*
$NdFeO_3$, *189, 190, 191*
Neurotransmitters sensor, *184, 189*
Non-enzymatic glucose sensor, *181, 182*
Properties, *178*
Sensor for aminoacids, *192*
Sensor for drugs, *188*
Sodium dodecyl sulfate (SDS), *189, 191*
$Sr_2Pd_{1-y}Au_yO_3$, *181*
Sr_2PdO_3, *181*
$Sr_{2-x}Ca_xPdO_3$, *182*
Applications of SAMs
Au-S bond, *125, 126, 128*
cellular membrane, *123*
Cholesterol sensors, *139, 141*
Cysteine (Cys), *125, 126*
Diabetes mellitus, *136*
DNA sensor, *132, 134*
Dopamine (DA), *124, 125*
Electroanalytical applications, *123*
Electrostatic interaction, *125*
enzymatic glucose sensors, *136*
Epinephrine (EP), *126, 127, 128*
Ferulic acid, *130*
Glucose oxidase enzyme (GOx), *137*
Glucose sensors, *136, 139*
Hydrogen peroxide sensors, *135, 136*
Immunosensor, *131, 132, 133*
Isoproterenol, *130*
long term stability, *125*
Metal sensors, *123*
molecule-molecule interaction, *128*
Neurotransmitter sensors, *124*
Non-enzymatic glucose sensors, *136*
Norepinephrine (NE), *129*
Peak separation, *126*
Properties, *122*
Sensor for ascorbic acid, *130*
Sensor for pharmaceutically important drugs, *129*
Sensor for uric acid, *130*
Sodium dodecyl sulfate (SDS), *125*
Substrate-molecule interaction, *128*

Assembly process
Alkanethiols, *105*, *106*
Chemisorption, *106*
Physisorption, *106*

B

Biosensors
Carbon nanotubes (CNTs), *83*
Carbon paste electrode (CPE), *84*
Cobalt phthalocyanine, *83*, *84*
Dobutamine, *84*
Dopamine (DA), *83*
Gold nanoparticles (Au), *84*
Iron(II) phthalocyanine, *83*
Morphine, *84*, *85*
Nickel phthalocyanine, *83*

C

Carbon based materials
Allotropes, *207*
Carbon, *207*, *208*
Sensors, *208*
Carbon paste (CP)
Graphite powder, *209*
Paraffin oil, *209*
Carboxylation and addition of organic groups
Adenine and guanine, *229*
Amide group, *226*
Atom transfer radical polymer (ATRP), *224*
Azomethine, *225*
Band gap, *223*
Benzoyl peroxide, *224*
Carboxylic acid, *227*
Covalent bond, *222*
Diazonium salts, *224*
DNA, *229*
Graphene oxide (GO), *226*, *227*
Graphene-polystyrene, *224*
Hydrazine hydrate, *228*
Nitrophenyls functionalization, *223*
Polymeric ionic liquid (PIL), *227*
Sulfonic acid functionalized graphene oxide (FSGO), *228*
Characterization of perovskites
BET, *173*
Calcination, *174*
DTG, *174*
EDAX, *169*
FTIR, *174*

Phases, *165*
SEM, *168*
TEM, *168, 170, 171, 172, 173*
Thermal analysis, *173*
XPS, *175, 176, 177*
XRD, *166, 167, 168, 173*
Characterization of SAMs
Atomic force microscopy, *120*
Contact angle goniometry, *122*
Electrochemical characterization, *115*
Electrochemical desorption, *115*
Electrochemical impedance spectroscopy, *115*
FT-IR, *118*
NMR, *121*
Oxidative desorption, *116, 117*
Raman spectroscopy, *118*
Reductive desorption, *116*
Scanning electron microscopy, *119*
Surface characterization, *117*
Transmission electron microscopy, *119*
UV-Vis, *121, 122*
XPS, *117*
Chemically Modified Electrodes, 51
Adsorption (Chemisorption), *53*
Applications, *52*
Approaches, *53*
Covalent bonding, *54*
Polymer film coating, *54*
Chemically Modified ELectrodes
Composite, *54*
Combined carbon materials
Carbon nanotubes (CNTs), *235, 237, 238*
Carbon paste electrode (CPE), *235, 237*
CVD, *237*
Cysteamine (CA), *236*
Folic acid (FA), *236*
Glutathione (Glu), *236*
Graphene oxide (GO), *237*
Graphene paste electrode, *236*
L-tryptophan, *235*
Mangiferin, *237*
NADH, *236*
N-hexyl-3-methylimidazolium hexafluorophosphate, *236*
Strain sensor, *238*
Cyclodextrins, *62*
Alpha, *64, 65*
Beta, *66, 67*
Gamma, *65*
Host-guest inclusion complex, *66, 67*
Structure, *62, 63*
Types, *63*

D

Doping of perovskites
    ABO_3 perovskites, *157, 158*
    A-site doping, *157, 158*
    B-site doping, *157, 158*
    Calcination conditions, *157*
    Doped perovskite, *157, 159*
    Non-enzymatic glucose sensor, *160*
    $Sr_{1.7}Ca_{0.3}PdO_3$, *159*
    $Sr_2Pd_{0.7}Au_{0.3}O_3$, *158*
    $SrTiO_3$, *159*
    synthesis method, *157*
Doping with foreign atoms
    Ag nanoparticles, *231*
    Bisphenol A (BPA), *231*
    B-N doped graphene, *234*
    Boron-doped graphene, *231*
    Chitosan (CS), *231*
    CVD, *230, 231*
    Density functional theory (DFT), *234*
    Dopamine (DA), *230*
    Exfoliation, *233*
    Glucose sensor, *231*
    N-doped graphene, *230*
    NO_2 gas sensor, *232*
    N-type semiconductor, *230*
    P-doped graphene, *233*
    P-type semiconductor, *230*
    S-doped graphene, *232, 233*
    S-N dual doped graphene (NS-G), *234*

E

Electrochemical sensors
    1-Butyl-1-methylpiperidinium hexafluorophosphate, *34*, 36
    1-Butyl-4-methylpyridinium tetrafluoroborate, *34*
    1-n-Hexyl-3-methyl imidazolium tetrafluoroborate, *34*
    Benazepril (BN), *34*
    Carbon paste electrode (CPE), *34, 36*
    Dopamine (DA), *36, 37, 38*
    Drug delivery, *42*
    Enoxacin (EN), *34*
    Ionic liquid crystals (ILCs), *34, 36*
    Ionic liquids, *40*
    L-dopa, *36, 38*
    Mediators, *51*
    NiO nanoparticles, *36, 37, 39*
    Paracetamol (ACOP), *36, 37, 38*
    Sodium dodecyl sulfate (SDS), *34, 37*
Examples of perovskites

A_2BO_3, *161*
ABO_3, *162*
$NdFeO_3$, *162*
Sr_2CuO_3, *161*
Sr_2PdO_3, *160, 161*

F

Factors affecting SAMs formation
  Chain length, *109, 110*
  Cysteine (Cys), *114*
  Macro and nano Au substrate, *113*
  Mixed SAMs, *110*
  Number of S-atoms, *111*
  Oxidative desorption, *114*
  Reductive desorption, *114*
  Repeated cycles stability, *114*
  Surface coverage, *112*
  Time of deposition, *112*
Ferrocene
  Bonding, *54, 55*
  Redox chemistry, *55*
  Structure, *54, 55, 57*
Functionalization
  18-crown-6, *217*
  Carbon paste electrode (CPE), *211, 212*
  Controlled potential, *209*
  Ferrocene, *215*
  Galvanostatic, *209*
  Gold nanoparticles (AuNPs), *212, 213*
  Mediator, *214*
  Metal nanoparticles, *212*
  Metal oxide nanoparticles, *212*
  Molecular imprinting polymer (MIP), *210, 211*
  Organic modifier, *214*
  Phthalocyanine, *215*
  Polymeric functionalization, *209*

G

General characteristics of perovskites
  Catalytic activity, *155, 156*
  Chemical stability, *156*
  Ferroelectricity, *155*
  Oxygen deficiency, *156, 157*
  Oxygen vacancies, *156*
  Piezoelectricity, *155*
  Superconductivity, *156*
Graphene
  2-dimentional material, *217*
  Applications, *217*

Graphene oxide (GO), *218*
Properties, *217*
Scotch-tape, *217*
SP^2 carbon, *217*
Surface area, *217*
Graphene functionalization
Covalent functionalization, *218*
Non-covalent functionalization, *218*
Graphite
Structure, *208*
Van der waal forces, *208*

H

Hydrogenation and halogenation of graphene
Fluorographene, *219*
Graphane, *219*
Hummer method, *220*
Hydrogen plasma, *219*
Hydrogen storage, *219*
Picric acid, *220*
XPS, *221*
Hydroxylation of graphene
Gas sensor, *222*
Graphene-OH, *221*

I

Ionic liquid crystals (ILCs)
General characteristics, *18*
Ionic liquids (ILs)
Characteristics, *17*

L

Liquid crystals, *19, 20*
Applications, 32
Classification, *21*
Lasers, *33*
LCD, *32*
Lyotropic, *27*
PDLC, *33*
Thermometers, *33*

M

Mesophases
Characterization, *27*
Differential scanning calorimetry, *30, 31*
Polarizing optical microscopy, *28, 29, 30*

X-ray diffraction, *31, 32*
Metallophthalocyanine
    Cobalt phthalocyanine, *86, 87*
    Iron(II) phthalocyanine, *88*
Models for perovskites
    Conduction band, *154*
    Covalent mixing, *155*
    Ionic model, *152*
    Madelung and electrostatic potential, *152, 153*
    Valence band, *154*

P

Perovskite oxides (ABO3)
    Characteristics, *151*
    Cubic structure, *149, 150*
    Hexagonal structure, *150*
    Orthorhombic structure, *150*
    Tolerance factor (t), *151*
Perovskite synthesis
    Combustion method, *163*
    Co-precipitation method, *164*
    Microwave synthesis, *164, 165*
    Pechini method, *163*
Phthalocyanine
    Applications, *81*
    Chemical structure, *80*
    Copper(II) phthalocyanine, *86*
    Derivatives, *81*
    Gas sensors, *84*
    Metallophthalocyanines (MPcs), *81, 82*
    Properties, *80*
    Redox reactions, *82*

S

SAMs preparation methods
    Chemical deposition, *107*
    Electrochemical deposition, *109*
    Gas phase deposition, *108*
    Solution deposition, *107*
Self-assembled monolayers (SAMs)
    Gold surface, *95, 96*
    Organothiols, *95, 96*
    Structure, *96*
Substrates for SAMs
    Ag-S bond, *103*
    Ascorbic acid (AA), *101*
    Copper (Cu), *102*
    Cysteine (Cys), *101*
    Dopamine (DA), *101*

Glassy carbon (GC), *101*
Gold (Au), *100, 101*
Mercury, *102*
Nickel (Ni), *103*
Palladium (Pd), *103, 104*
Platinum (Pt), *101*
S-Au bond, *101*
Silicon (Si), *105*
Silver (Ag), *102*
Zinc (Zn), *104*

T

Thermotropic liquid crystals, *21*
    Bent-core mesogens, *27*
    Calamitic mesogens, *21*
    Chiral phases, *24*
    Discotic mesogens, *26*
    Nematic phase, *22, 23*
    Smectic phase, *23*
Types of SAMs
    Carbenes and diazonium salts, *100*
    Organothiols, *97, 99*
    Phosphates, *99*
    Phthalocyanines, *99*
    Silanes, *100*

www.ingramcontent.com/pod-product-compliance
Lightning Source LLC
Chambersburg PA
CBHW050456190326
41458CB00005B/1309